DINOSAURS
THE GRAND TOUR

EVERYTHING WORTH KNOWING ABOUT DINOSAURS
FROM AARDONYX TO ZUNICERATOPS

Keiron Pim

with field notes by Jack Horner
Illustrated by Fabio Pastori

THE EXPERIMENT

DINOSAURS – THE GRAND TOUR: *Everything Worth Knowing About Dinosaurs from Aardonyx to Zuniceratops*

Copyright © Keiron Pim, 2013
Text on pages 106–109, 156–158, 184–189 © The Experiment, 2014
Illustrations © Fabio Pastori, 2013

First published in Great Britain as *The Bumper Book of Dinosaurs* by Square Peg/Random House, 2013

The Experiment, LLC
220 East 23rd Street, Suite 301
New York, NY 10010-4674
www.theexperimentpublishing.com

The Experiment's books are available at special discounts when purchased in bulk for premiums and sales promotions as well as for fund-raising or educational use. For details, contact us at info@theexperimentpublishing.com.

Library of Congress Cataloging-in-Publication Data

Pim, Keiron., author
  Dinosaurs--the grand tour : everything worth knowing about dinosaurs from Aardonyx to Zuniceratops / by Keiron Pim.
      pages cm
  First published in Great Britain as The Bumper book of dinosaurs by Square Peg, 2013.
  Includes index.
  ISBN 978-1-61519-212-0 (hardcover) -- ISBN 978-1-61519-213-7 (ebook) 1. Dinosaurs.
2. Dinosaurs--Pictorial works. I. Title.

  QE861.4.P56 2014
  567.9--dc23

2014018581

ISBN 978-1-61519-212-0
Ebook ISBN 978-1-61519-213-7

Distributed by Workman Publishing Company, Inc.

Cover design by Orlando Adiao
Author photos of Keiron Pim and Jack Horner © Keiron Tovell and the Museum of the Rockies / Kelly Gorham, respectively
Photo on page 109 courtesy the Museum of the Rockies
Designed and produced by The Curved House
Art Direction: Kristen Harrison
Design: Rowan Powell
Artworking: Sophie Devine
Layout and typesetting in DINPro: Jonathan Baker
Additional typesetting by Neuwirth & Associates, Ltd.
Manufactured in China

First printing August 2014
10 9 8 7 6 5 4 3 2 1

For Isla, Lottie and Rosa

DROMAEOSAURUS   DEINONYCHUS   "MEGARAPTOR"   TRO

SCIPIONIX SAMNITICUS   VELOCIRAPTOR   OVIRAPTOR

UNENLAGIA   CONFUCIUSORNIS SANCTUS   ARGENTAVIS MAGNIFICENT

OSAUROPTERIX   PROTARCHAEOPTERIX   CAUDIPTERIX   ARCHAEOPTERIX

INTRODUCTION

If you think you know the dinosaurs' world, then think again, for it grows stranger and more fascinating all the time. We are living in the 'golden age' of dinosaur discovery, with dozens of significant fossils emerging every year: where once our picture of the Mesozoic era was grey and populated by lumpen reptiles, now it is gaining colour and sound, revealing in ever-greater detail the most dynamic and spectacular creatures to have existed.

For decades, dinosaurs have roamed around young children's imaginations, much as they once wandered the Earth. Now, the increasingly vivid image painted by this deluge of discoveries is enticing many adults and teenagers into rediscovering the thrill of the prehistoric world, rekindling that tingle of wonder that they felt on first stepping into London's Natural History Museum and pausing to gaze up at the overwhelming *Diplodocus* skeleton. This is a book with plenty to educate, entertain, amuse and delight the reader across all ages: child, parent and grandparent alike.

For as well as being extraordinary in their own right, dinosaurs can teach us much about the modern world. Read about them here and you will find yourself easily absorbing aspects of so many other topics: geology, history, religion, evolutionary theory, anatomy, astronomy, even Native American and Chinese mythology. Investigating these amazing creatures also makes us think harder about the diversity of life that has existed over many millions of years on this planet.

Some dinosaurs provoke a similar thrill to watching a horror film (think for a moment about *Tyrannosaurus* and its 1.5m (5ft) jaws packed with 23cm (9in) teeth and try not to shiver) while others are just plain comical – or at least their names are. If you want to find out more about *Irritator*, are intrigued by the Lewis Carroll reference in the name of *Borogovia*, or want to know why *Dracorex hogwartsia* got its name and why it might lose it again, then this is the book for you. More seriously, trying to comprehend how long ago they lived and thus dwelling on the mindboggling expanse of geological 'deep time' prompts us to marvel on another level.

In scientific terms dinosaurs are a sub-group of the land-living archosaur group of reptiles in which the legs are directed straight down underneath the body, rather than bowing outwards. From *Aardonyx*, a lumbering beast that formed a link between two-legged and four-legged dinosaurs, through to *Zuniceratops*, possessor of a deadly pair of horns, the 300-plus known varieties encompassed huge diversity. Not all of the strange beasts alive during their reign were dinosaurs,

of course: it is easy to forget that they lived alongside mammals, lizards and birds that looked similar to their modern equivalents – as well as some amazing extinct animals quite unlike anything living today. Creatures such as the plesiosaurs, ichthyosaurs and pterosaurs are sometimes lazily termed dinosaurs – which they weren't, but are still fascinating, so we will look at some of them too. It would be remiss to explore this era and leave out on a technicality *Quetzalcoatlus*, a gigantic, stork-like flying reptile with a 10m (33ft) wingspan.

As well as looking at dinosaurs of all shapes and sizes, from the immense *Argentinosaurus* to the minuscule *Microraptor*, we will travel through their habitats – forests, arid scrubland, swamps and coastal plains, baking deserts to the polar regions – and see how much the planet has changed since they were here. Back in the Triassic the landmasses were more connected and a *Thecodontosaurus* could plod its way from modern Australia to Antarctica; over many millions of years the land fractured and drifted into the arrangement of continents that we know today. In 120m years' time, according to one prediction, the world will appear very different again, with North and South America having separated, and Africa rubbing up against Britain and Western Europe.

The much–debated issue of the dinosaurs' extinction remains unresolved, though many palaeontologists have a broad consensus on the likeliest causes. This book looks at the latest ideas on the subject. But a more interesting question to ask might be: are the dinosaurs extinct? For while it is awe-inspiring to consider how long ago they lived, it is even more thrilling to find the evidence of their 'lost world' that surrounds us today. As well as exploring the best places to go fossil-hunting, we will examine some curious continuations of those ancient times still flourishing: on the one hand the so-called 'living fossils' such as the coelacanths that have somehow endured from then until now, but also genuine dinosaurs that we all see daily. If you want to know the link between a *T. rex* and a roast dinner, then you'll find it here.

We hope that this book will cause you to dwell upon the wonders of the prehistoric world – but also to get out and explore the beauty of our planet today. From bones emerging out of crumbling cliffs and quarries not so far from where you live, to the birds that we see flying overhead, we can find dinosaurs all around us if we only know where to look.

NOTE ON GEOLOGICAL ERAS

When geologists and palaeontologists refer to periods such as the Jurassic and Cretaceous, they generally refer to the Early, Middle or Late sub-divisions. When referring to the rock beds dating from those periods, they use the terms Lower, Middle and Upper. So for instance *Allosaurus* lived in the Late Jurassic, but its fossils have been found in the Upper Jurassic sandstone beds of North America's Morrison Formation.

Overleaf are the main geological periods of Earth's history – and a handy way of memorising them.

first bacteria, algae, worms, jellyfish

trilobites, shellfish, sponges, segmented worms

jawless fish, first vertebrates

cartilaginous fish, first plants growing on land, marine invertebrates

bony fish, first ammonites, vascular plants

amphibians, first reptiles, first insects

mammal-like reptiles, sometimes known as protomammals

PRECAMBRIAN
4500MYA

CAMBRIAN
542MYA

ORDOVICIAN
488MYA

SILURIAN
443MYA

DEVONIAN
416MYA

CARBONIFEROUS
359MYA

PERMIAN
299MYA

IF YOU CAN'T REMEMBER THE ERAS, TRY THIS MNEMONIC: PREGNANT CAMELS ORDINARILY SIT DOWN CAREFULLY. PERHAPS THEIR JOINTS CREAK!

first dinosaurs, turtles, lizards, crocodiles and first mammals

dinosaurs, first birds

first flowering plants, the peak of the dinosaurs' reign until their extinction

TRIASSIC
251MYA

JURASSIC
199MYA

CRETACEOUS
145-65MYA

EARLY   MIDDLE   LATE

EARLY   MIDDLE   LATE

EARLY   LATE

9

Dinosaurs may be divided into two main groups and several sub-groups. They include:

Ornithischians

dinosaurs that had a four-pronged pelvis with the pubis bones pointing backwards towards the tail, resembling that of modern birds – though birds actually descended from saurischians. Ornithischians included...

Ceratopsians

small to medium-sized four-legged herbivores that had a variety of facial horns and usually a neck frill.

Ankylosaurs

small to medium-sized herbivores with an armour-plated back, a spiky tail and often spikes protecting the neck.

Stegosaurs

medium-sized herbivores with a small head, high plates or spikes along the back and a spiky tail.

Pachycephalosaurs

small to medium-sized herbivores with greatly thickened skull-bones and often small horns on the head.

Ornithopods

small to large-sized herbivores that generally walked on two legs and had birdlike feet, hence their name. The most significant group is the...

Iguanodontians

large herbivores ranging from the well-known *Iguanodon* to the later duck-billed hadrosaurs.

Saurischians

dinosaurs that had a three-pronged pelvis with the pubis bone pointing forwards, resembling that of modern lizards. Saurischians included...

Sauropods

large to gigantic herbivores that walked on four legs and typically had a bulky body, long tail, long neck and small head. Notable types included...

Diplodocids

very long and relatively slender sauropods with especially small heads and whip-like tails.

Brachiosaurs

extremely tall sauropods with a giraffe-like posture.

Titanosaurs

some of the hugest, bulkiest land animals ever known, often more than 30m (100ft) long and sometimes weighing more than 90 tonnes (88 tons).

Theropods

two-legged, mostly carnivorous animals that included...

Abelisaurs

medium-sized speedy predators with tiny arms and a powerful tail.

Megalosaurs

medium to large-sized carnivores, including...

Spinosaurs

medium to large-sized crocodile-skulled fish-eaters.

Allosaurs

medium to large-sized carnivores, which included...

Carcharodontosaurs

very large carnivores with shark-like teeth.

Alvarezsaurs

small, feathered insectivores with tiny hands sometimes reduced to a single finger.

Deinonychosaurs

small to medium-sized birdlike predators, which included...

Dromaeosaurs

fierce hunters with a retractable sickle-shaped claw on each foot.

Troodontids

omnivores with a smaller sickle-claw and very advanced hearing.

Ornithomimosaurs

ostrich-like, omnivorous running dinosaurs.

Oviraptorosaurs

beaked, toothless, feathered omnivores.

Therizinosaurs

medium-to-large feathered herbivores with huge claws.

Tyrannosaurs

medium-to-large carnivores with heavy skulls, thickened teeth and small arms.

WHILE WE CAN REFER TO DINOSAURS GENERALLY AS BEING ABELISAURS, STEGOSAURS, TYRANNOSAURS AND SO ON, TO BE PRECISE PALAEONTOLOGISTS SPEAK OF, FOR INSTANCE, ABELISAURIDS AND ABELISAUROIDS – THE FORMER BEING THOSE WITHIN THE ABELISAURIDAE FAMILY, AND THE LATTER BEING THOSE THAT WERE EVIDENTLY RELATED BUT SIT JUST OUTSIDE IT.

KEY

Symbols

Each dinosaur in this book is accompanied by a list of symbols to help you quickly understand its characteristics.

Length

The dinosaur's estimated length from head to tail.

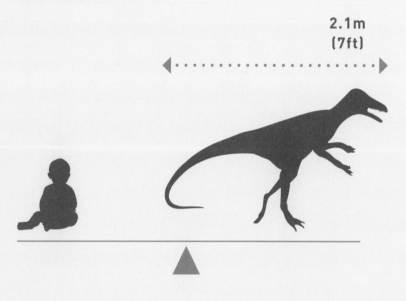

2.1m
(7ft)

JURASSIC

TITHONIAN	
KIMMERIDGIAN	LATE
OXFORDIAN	
CALLOVIAN	
BATHONIAN	MIDDLE
BAJOCIAN	
AALENIAN	
TOARCIAN	
PLIENSBACHIAN	EARLY
SINEMURIAN	
HETTANGIAN	

150mya

C
CARNIVOROUS

750kg
(1600lb)

ARGENTINA

Timeline

The period when the dinosaur lived, e.g. Jurassic, the epoch within that period, e.g Late Jurassic, and the age within that epoch, e.g. Kimmeridgian. Each age itself represents several million years, and the corresponding rock layer is called a stage. Most are named after places where rocks from those times were found, e.g. the village of Kimmeridge in Dorset, or Pliensbach in Germany.

Clock

How many million years ago (mya) the dinosaur lived. Where very long timespans are given – e.g. 215–200mya – this refers to the approximate age of the rocks containing the fossils, rather than how long the species existed.

Diet

Whether the dinosaur was carnivorous (eating meat), herbivorous (eating plants), omnivorous (eating meat and plants), insectivorous (eating insects) or piscivorous (eating fish).

Weight

The dinosaur's estimated weight in metric and imperial measurements. A metric tonne is 1000kg, so a 50,000kg sauropod was 50 tonnes (49.2 imperial tons).

Geography

Where the dinosaur's fossils were found in modern-day countries. See pages 284-285 to learn more about how our continents have shifted and changed over time.

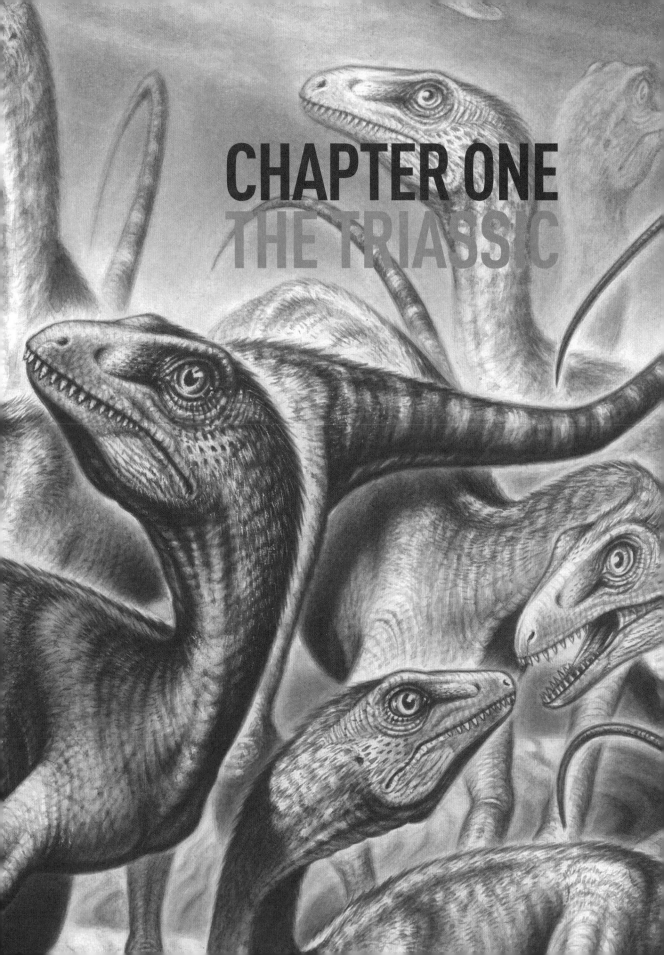

CHAPTER ONE
THE TRIASSIC

THE TRIASSIC

It began around 250mya with the Permian–Triassic extinction event, the so-called 'Great Dying' that mysteriously obliterated nine out of ten marine animals and seven out of ten land-living species, almost extinguishing life on Earth... but relatively soon afterwards the dinosaurs began to rise. Around 230mya, halfway through the Triassic period, they evolved from among a group of small, slim-limbed archosaurian predecessors and commenced an incredible 160m-year era in which they would dominate the Earth.

The dinosaurs' early evolution is gradually coming to light now: while peering back into the darkest depths of geological time is a difficult business because there are relatively few well-preserved remains, with every good find yielded by fossil-rich formations such as Argentina's Ischigualasto National Park (better known as Valley of the Moon owing to its lunar clay landscape), a more detailed understanding develops of the emergence of dinosaur life. This was an era in which today's continents were connected in one great landmass called Pangaea, which was only habitable around its periphery; the centre was a scorching desert, so far from the surrounding ocean (called Panthalassa) that moisture never reached it. But around the cooler fringes of that huge tract of land an astonishing diversity of animals arose, of which dinosaurs at first only formed a small minority. Mammals began to evolve, winged

flying reptiles known as pterosaurs
flew through the skies, long-necked
plesiosaurs and fish-like ichthyosaurs
roamed the seas... and then gradually,
from being primitive little omnivorous
hunters, the dinosaurs began to
diverge into carnivorous theropods
and herbivorous ornithischians and
prosauropods. Here were the distant
ancestors of the terrible killers and
gargantuan plant-eaters that would
represent the dinosaurs' apex more
than 100m years later.

The era closed with another great
series of extinctions over millions
of years, but the dinosaurs were
not hindered; indeed the absence
of former rivals enabled them to
flourish. By the end of the Triassic,
they were ready to rule the world.

RHAETIAN	LATE
NORIAN	
CARNIAN	
LADINIAN	MID
ANISIAN	
OLENEKIAN	EARLY
INDUAN	

(her-AYR-ah-SORE-us)

HERRERASAURUS ISCHIGUALASTENSIS ········· AND ·······

231mya

C

CARNIVOROUS

350kg
(772lb)

ARGENTINA

Ever since 1959 when a goatherd named Victorino Herrera noticed its bones in the foothills of the Andes, this carnivorous biped with a deep head and huge teeth has prompted great debate. Some doubted whether it was a true dinosaur, with an alternative opinion being that it was a sort of ancestral, proto-dinosaur. If it was a dinosaur, questions surrounded whether it was a theropod or sauropodomorph. The discovery in 1988 of a near-complete fossil including a skull helped clarify matters and most (though not all) experts agree that *Herrerasaurus* and its relatives are early theropods.

up to 6m
(20ft)

TRIASSIC

RHAETIAN	
NORIAN	LATE
CARNIAN	
LADINIAN	MID
ANISIAN	
OLENEKIAN	EARLY
INDUAN	

(san-hwan-SORE-us)

SANJUANSAURUS GORDILLOI

4m
(13ft)

231mya

C

CARNIVOROUS

240kg
(550lb)

ARGENTINA

This shows that at the very dawn of confirmed dinosaurs' presence in the fossil record circa 230mya, the three main types already existed: ornithischians (bird-hipped dinosaurs), and within the saurischians (lizard-hipped dinosaurs) the sauropodomorphs and carnivorous theropods. That is why we can't talk about any one species being 'the first dinosaur'. While a study in 2012 of the 243-million-year-old *Nyasasaurus* suggests it was the earliest known dinosaur, this remains unconfirmed, and other experts would prefer to term it a dinosauriform, a closely related animal just outside the Dinosauria proper. The patchy evidence we have from ancient rocks shows no confirmed dinosaurs, and then suddenly several kinds in each of these categories.

Herrerasaurus gives its name to the Herrerasauridae group, which also includes *Staurikosaurus* (page 24) and the recently discovered *Sanjuansaurus gordilloi*. *Sanjuansaurus* was discovered in Argentine rocks of the same age, in San Juan province naturally enough, and was very similar to a medium-sized *Herrerasaurus*. Its fossils, discovered in 1994, included long legbones that suggest *Sanjuansaurus* was the fastest runner in this family of very early hunters.

19

TRIASSIC

RHAETIAN	
NORIAN	LATE
CARNIAN	
LADINIAN	MID
ANISIAN	
OLENEKIAN	EARLY
INDUAN	

(cro-mo-gi-SORE-us)

CHROMOGISAURUS NOVASI

231mya

? OMNIVOROUS?

uncertain

SAN JUAN, ARGENTINA

20

MEANING 'COLOURED EARTH DINOSAUR'

These mysterious creatures were primitive sauropodomorphs, dinosaurs beginning to assume the form of the later sauropods such as *Diplodocus*, and they were found in the beautifully stratified Valle Pintado of Argentina, hence *Chromogisaurus* meaning 'coloured earth dinosaur'. *Panphagia* means 'eats everything', as its jaws and teeth suggest it could bite flesh and chew plants – which is interesting given that in the Jurassic and Cretaceous its distant descendants would evolve into highly specialised huge herbivores that reached such enormous sizes by constantly eating vegetation. *Panphagia* thus marks a transition between the early hunters and the plant-eaters. *Chromogisaurus* may well also have been an omnivore, but its skull has not been found so it's currently impossible to be sure.

2m
(7ft)

TRIASSIC

RHAETIAN	
NORIAN	LATE
CARNIAN	
LADINIAN	MID
ANISIAN	
OLENEKIAN	EARLY
INDUAN	

AND ·······

(pan-FAYJ-ee-ah)

PANPHAGIA PROTOS

1m (3ft)

231mya

0

OMNIVOROUS

10kg
(22lb)

SAN JUAN,
ARGENTINA

Panphagia's discovery was announced in 2009, three years after its reddish-coloured fossils were found in the same grey-green sandstone stratum as the slightly larger *Chromogisaurus*. The Ischigualasto Formation amounts to the clearest window we have into the Late Triassic, and in recent years has offered repeated flashes of light into the era's murkiest depths. According to Martin Daniel Ezcurra, the South American palaeontologist who named it in 2010, *Chromogisaurus*' discovery made it clearer to palaeontologists just how diverse dinosaurs already were when they appear in the fossil record. However, while they would proceed to dominate the world, at this stage in South America they formed only around 6 per cent of the animals known to have existed.

TRIASSIC

RHAETIAN	
NORIAN	LATE
CARNIAN	
LADINIAN	MID
ANISIAN	
OLENEKIAN	EARLY
INDUAN	

231mya

C

CARNIVOROUS,
SMALL
REPTILES

10kg
(22lb)

NORTH-
WESTERN
ARGENTINA

22

(EE-oh-rap-tor)

EORAPTOR LUNENSIS · · · · · · · AND · · ·

MEANING
'DAWN
PLUNDERER'

These primitive and little-understood creatures stalked the riverbanks of Late Triassic Argentina, perhaps hunting small reptiles. *Eoraptor* was regarded for a time as the earliest known theropod, presumably the ancestor of all the great meat-eaters that followed. It was this idea that caused American palaeontologist Paul Sereno and colleagues to give it a name meaning 'dawn plunderer' in 1993. But in 2011 further research caused them to reassess its position. The discovery of a longer, leaner predator called *Eodromaeus* ('dawn runner') from the same Ischigualasto Formation rocks revealed that *Eoraptor* had a contemporary with far more pronounced theropod features. These included serrated, sabre-shaped teeth; long-fingered grasping hands with sharp claws; long legs for fast running (up to 19mph, it is surmised); and a stiffened tail that enhanced its balance, enabling swift turns while running. *Eoraptor* lacks these features, and has characteristics that would later appear in the herbivorous sauropodomorphs, such as enlarged nostrils and an inset lower first tooth. Despite the fact that it was probably a carnivore itself, Sereno's team removed *Eoraptor* from the theropods and moved it to the base of the sauropodomorph family tree.

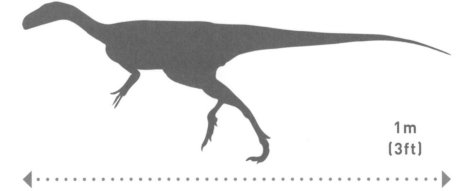

1m
(3ft)

(EE-oh-DROME-ee-us)
EODROMAEUS MURPHI

MEANING 'DAWN RUNNER'

TRIASSIC

RHAETIAN	LATE
NORIAN	
CARNIAN	
LADINIAN	MID
ANISIAN	
OLENEKIAN	EARLY
INDUAN	

231mya

C

CARNIVOROUS, SMALL REPTILES

5kg (11lb)

NORTH-WESTERN ARGENTINA

1.2m (3ft 11in)

23

TRIASSIC

RHAETIAN		
NORIAN	LATE	
CARNIAN		
LADINIAN	MID	
ANISIAN		
OLENEKIAN	EARLY	
INDUAN		

227–221 mya

C

CARNIVOROUS

12kg (26lb)

RIO GRANDE DO SUL, SOUTHERN BRAZIL

24

(STORE-ick-oh-SORE-us)

STAURIKOSAURUS PRICEI

MEANING 'SOUTHERN CROSS LIZARD'

This primitive predator was named in 1970, more than 30 years after its bones were uncovered in Rio Grande do Sul by a Brazilian palaeontologist with a somewhat surprising name: Llewellyn Ivor Price, whose parents were Americans of Welsh stock. At that stage it was rare to find dinosaur remains south of the equator, hence the name meaning 'Southern Cross lizard'. The Brazilian national flag features the four stars that form that constellation, which is only visible from the southern hemisphere. *Staurikosaurus*' long hind legs gave it great pace for pursuing its prey.

2.1m (7ft)

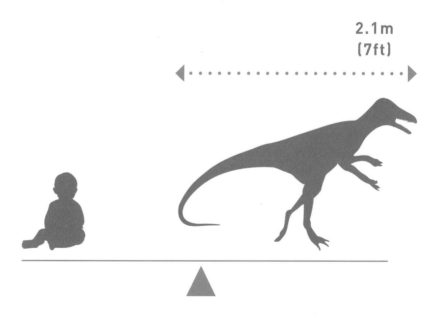

(CHIN-dee-SORE-us)
CHINDESAURUS BRYANSMALLI

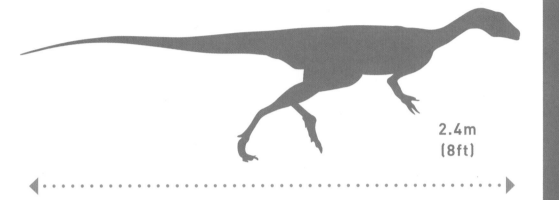

2.4m
(8ft)

TRIASSIC

RHAETIAN	
NORIAN	LATE
CARNIAN	
LADINIAN	MID
ANISIAN	
OLENEKIAN	EARLY
INDUAN	

227–210
mya

CARNIVOROUS

15kg
(33lb)

This is possibly the earliest American dinosaur. It had long legs, a whip-like tail and, it is assumed, a deep and narrow skull. Several partially preserved specimens have now been found, one of them featuring a single serrated tooth – which would seem to confirm *Chindesaurus* as a carnivore, but a recent analysis suggests that the tooth came from a different dinosaur.

Pinpointing Triassic dinosaurs' position near the base of the family tree is tricky work at any time, but especially when only fragmentary remains are available. After fossil-hunter Bryan Small unearthed the first known *Chindesaurus* bones at Chinde Point, Petrified Forest National Park, in 1985 it was considered a herbivorous 'prosauropod' – an informal, catch-all term for those sauropodomorphs that aren't sauropods – but now this and its fellow herrerasaurids are generally considered to have been theropods.

POSSIBLY THE EARLIEST AMERICAN DINOSAUR

ARIZONA, NEW MEXICO AND TEXAS, USA

25

TRIASSIC

RHAETIAN	
NORIAN	LATE
CARNIAN	
LADINIAN	MID
ANISIAN	
OLENEKIAN	EARLY
INDUAN	

205mya

C

CARNIVOROUS

200kg
(440lb)

GERMANY

(lil-ee-en-STERN-us)

◄···· *LILIENSTERNUS LILIENSTERNI*

Few of the early dinosaurs attained a great size but at 5.2m (17 ft) long *Liliensternus* posed a formidable threat to Late Triassic Germany's prosauropods, such as *Plateosaurus*. One of the foremost killers of the early dinosaur era, this was a slender, long-necked and long-tailed creature with four fingers on its hand – and the fourth was smaller than the rest, paving the way for the three-fingered hands of later carnivores. Its head sported a long, probably brightly coloured crest that could have served for display or communication purposes.

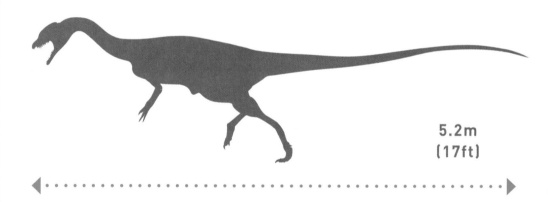

5.2m
(17ft)

LONGER
THAN A
BIG CAR!

TRIASSIC

RHAETIAN	
NORIAN	LATE
CARNIAN	
LADINIAN	MID
ANISIAN	
OLENEKIAN	EARLY
INDUAN	

(bli-KAHN-oh-SORE-us)

BLIKANASAURUS CROMPTONI

It's likely this was an early sauropod, and it's definitely the earliest dinosaur known to have spent all its time on all-fours, a fact deduced from a single massive hind left leg with the short foot characteristic of quadrupeds. The proportions of that fossil limb, found in Lower Elliot Formation Rocks, suggest a heavily built animal. It was named in 1985 after Mount Blikana in Lesotho, southern Africa, where the fossil was found in 1962.

216–203 mya

HERBIVOROUS, HIGH BROWSER

250kg (550lb)

LERIBE, LESOTHO

5m (16ft)

TRIASSIC

RHAETIAN	
NORIAN	LATE
CARNIAN	
LADINIAN	MID
ANISIAN	
OLENEKIAN	EARLY
INDUAN	

228–216
mya

H

**HERBIVOROUS,
LOW BROWSER**

2kg
(5lb)

**ISCHIGUALASTO,
NORTHERN
ARGENTINA**

(pee-SAHN-o-sore-us)

PISANOSAURUS MERTI

This perplexing dinosaur had jaws that have prompted confusion since its discovery in 1967. At present it is viewed as the oldest known ornithischian dinosaur. Its head is certainly like an ornithischian's, as it has its characteristic cheeks and method of chewing – but its body is more like a saurischian's, in particular the pelvis and ankle joints. The fossil was far from complete, which makes assessing its stature very hard, but it would have been a little biped nipping through the lush, well-watered conifer forests that covered Argentina at the time. *Pisanosaurus* lived among many types of primitive reptile, such as the rhynchosaurs and rauisuchians, but there were a few other dinosaurs here too. Among them was *Herrerasaurus*, which would have counted among *Pisanosaurus'* predators.

**LOW
BROWSING**

1.3m
(4.5ft)

(PLAT-ee-oh-SORE-us)

PLATEOSAURUS ENGELHARDTI

TRIASSIC

RHAETIAN	LATE
NORIAN	
CARNIAN	
LADINIAN	MID
ANISIAN	
OLENEKIAN	EARLY
INDUAN	

Other species: *P. gracilis*

up to 10m
(33ft)

205mya

H

HERBIVOROUS, HIGH BROWSER

4000kg
(3.9 tons)

GERMANY, SWITZERLAND AND FRANCE

More than 100 *Plateosaurus* skeletons have been found across central and western Europe, making this big biped one of the best known of all European dinosaurs. Dozens have been found in single locations, which suggests they were herding animals. Hermann von Meyer named it in 1837 and it was the fifth dinosaur recognised that remains so today – that is to say, forgetting about other early 'discoveries' that have proved invalid. During the 175 years since its discovery experts considered it first as an upright biped, then as walking on all-fours, but the current picture is of a biped that walked with a horizontal back and tail, a posture that fits with the fact that its forelimbs were half the length of its hind limbs. This all serves to remind us just how drastically our notions of dinosaurs' appearance and behaviour can alter as new evidence emerges or existing evidence is more rigorously analysed. *Plateosaurus* was one of the prosauropods, that informal grouping of mostly Triassic herbivores that stretched into the Early-to-Middle Jurassic. The unusually wide variation between fully grown adults' sizes, which span from 5m (16ft) to 10m (33ft), helps explain why numerous *Plateosaurus* species have been named over the years. Nowadays only two are accepted.

TRIASSIC

RHAETIAN	
NORIAN	LATE
CARNIAN	
LADINIAN	MID
ANISIAN	
OLENEKIAN	EARLY
INDUAN	

(see-LOF-ih-sis)
COELOPHYSIS BAURI

200mya

C

CARNIVOROUS,
SMALL HUNTER

25kg
(55lb)

NEW MEXICO,
USA

This lithe running dinosaur is the best known meat-eater of the Late Triassic – hundreds of complete or near-complete fossils have been discovered. Edward Drinker Cope found the first in 1889 during the Bone Wars (see page 88) and gave it a name meaning 'hollow form', referring to a characteristic that its bones share with those of modern birds. Something else they had in common was a wishbone – *Coelophysis* is the earliest dinosaur proved to possess one. The most famous collection of *Coelophysis* fossils was found in 1947 at Ghost Ranch, New Mexico, when Edwin Colbert's team stumbled upon a densely packed dinosaur graveyard. Within a 6m by 20m (20ft by 65ft) area lay the bones of hundreds of *Coelophysis*. A few skeletons' stomachs contained the bones of smaller reptiles, which looked similar to *Coelophysis* but were too large to be embryos – so Colbert's theory was that adults were cannibalistic and preyed on their young. Later the bones were actually shown to belong to little crocodilians they had eaten. But mystery continues to shroud why so many of the *Coelophysis* were found together in such excellent condition. There was no evidence of their being attacked or swept to a single deposit point by a flash flood. They just appear all to have died suddenly and been covered before scavengers could disturb them. Colbert suggested that volcanic gases overcame them. The theory has no supporting evidence, but neither has anyone else managed to provide a better answer.

HOLLOW
BONES AND
WISHBONE AS
IN BIRDS

3m
(10ft)

(DEE-mon-oh-SORE-us)

DAEMONOSAURUS CHAULIODUS

TRIASSIC

RHAETIAN	LATE
NORIAN	
CARNIAN	
LADINIAN	MID
ANISIAN	
OLENEKIAN	EARLY
INDUAN	

1.5m
(5ft)

205mya

CARNIVOROUS

15kg
(33lb)

NEW MEXICO,
USA

Many Triassic carnivores looked very similar – long slender skull, jaws filled with small sharp teeth, a lean, long-legged body – so *Daemonosaurus'* discovery in 2011 served to shake up preconceptions about early theropods. Unearthed at Ghost Ranch Quarry like the more typically Triassic *Coelophysis*, the crushed and partial remains included a short and deep skull containing large, protruding teeth. The species name means 'buck-toothed', which makes *Daemonosaurus* sound goofier than it must really have been. With its serrated fangs and tough jaws, it had a powerful and penetrating bite. However, it does not appear to have been an evolutionary success story: the researchers who described it consider *Daemonosaurus* a dead-end, with no connection to the bigger, better known short-snouted theropods of the Jurassic and Cretaceous. The demonic reference in its name does not refer to the dinosaur's appearance, incidentally, but to mythology that evil spirits haunt Ghost Ranch, a barren expanse of badlands in the heart of New Mexico.

NICKNAMED THE 'BUCK-TOOTHED DINOSAUR'

31

TRIASSIC

RHAETIAN	
NORIAN	LATE
CARNIAN	
LADINIAN	MID
ANISIAN	
OLENEKIAN	EARLY
INDUAN	

(REE-oh-hah-SORE-us)
RIOJASAURUS INCERTUS

210mya

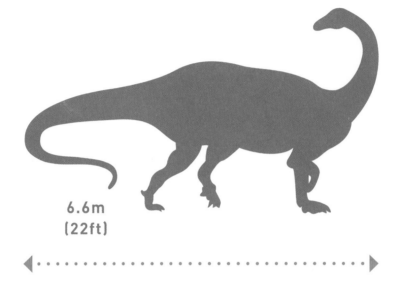

6.6m
(22ft)

HERBIVOROUS, HIGH BROWSER

800kg (1800lb)

Big prosauropods such as *Riojasaurus* look very similar to the true sauropods – in fact this dinosaur is thought to have been the heaviest land animal that existed before their appearance. Argentine palaeontologist Jose Bonaparte discovered it in the Andean foothills during the 1960s. The specimen had a long neck and no skull but he predicted that its head must have been small, and a later find proved him right. Along with its close relative *Eucnemesaurus*, unearthed in what is now South Africa (which in the Triassic connected with the eastern side of South America), it forms the family Riojasauridae.

RIOJA PROVINCE, ARGENTINA

LESSEMSAURUS SAUROPOIDES ·········▶

TRIASSIC

RHAETIAN	LATE
NORIAN	LATE
CARNIAN	
LADINIAN	MID
ANISIAN	MID
OLENEKIAN	EARLY
INDUAN	EARLY

All that is known of this dinosaur is a vertebral column but this revealed something rare among Triassic dinosaurs: a series of spines that formed a ridge running down *Lessemsaurus'* back. As with similar features on later dinosaurs, its purpose remains debated: it may have helped regulate body temperature or been for display. *Lessemsaurus* lived alongside *Riojasaurus* in the wet woodlands of Triassic Argentina. Jose Bonaparte named this early sauropod in 1999 for Don Lessem, an American popular science author nicknamed 'Dino Don', as his books often feature dinosaurs.

210mya

HIGH BROWSING

H

HERBIVOROUS, HIGH BROWSER

1800kg (1.7 tons)

9m (30ft)

RIOJA PROVINCE, ARGENTINA

TRIASSIC		
RHAETIAN	LATE	
NORIAN		
CARNIAN		
LADINIAN	MID	
ANISIAN		
OLENEKIAN	EARLY	
INDUAN		

210mya

H

HERBIVOROUS,
LOW BROWSER

900kg
(1980lb)

GERMANY

(ef-RAAS-ee-ah)

EFRAASIA DIAGNOSTICUS

Germany has the world's best concentration of Triassic prosauropod fossils and this typical example was named in 1973 in honour of Eberhard Fraas, who found its remains in 1909 and did much to bring the country's wealth of ancient relics to light. *Efraasia* was first deemed a fairly small animal but closer analysis revealed the skeleton to be that of a young dinosaur. The fact that its bones were fossilised alongside a crocodilian jaw added confusion for a while, leading to *Efraasia*'s depiction as a carnivore. Current thinking is that it walked on all fours but could rear up on its hind legs to feed, grasping vegetation with its long fingers.

Fraas thought that his discovery represented a new species of *Thecodontosaurus* but died before he could finish describing it. This prompted a complicated tangle of classifications over the subsequent 60 years: at various times it has been placed in genera such as *Teratosaurus*, *Paleosaurus*, *Palaeosauriscus* and *Sellosaurus*, but for now it is considered a dinosaur in its own right.

6m
(20ft)

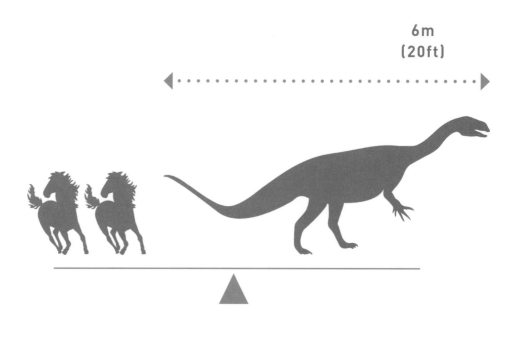

(pan-tee-DRAY-coe)

PANTYDRACO CADUCUS

1.8m (6ft)

It is hard to imagine but this was a distant ancestor of the great sauropods such as *Apatosaurus*, who lived 50m years later. Whereas those heavily built animals thundered along on all fours, prosauropods such as *Pantydraco* were semiquadrupedal – that is, they walked on four legs but often reared up on their hindquarters. This made them the first herbivores that could browse leaves from high branches. Estimates of its adult size are sketchy as the fossil remains, discovered in south Wales, are only partial and from a young animal. *Pantydraco* was for a while considered a species of *Thecodontosaurus* but further study in 2007 proved it merited its own genus. The name refers to the Pant-y-ffynnon quarry where its remains were found – and the 'draco' comes from the Latin for 'dragon'.

TRIASSIC

RHAETIAN	
NORIAN	LATE
CARNIAN	
LADINIAN	MID
ANISIAN	
OLENEKIAN	EARLY
INDUAN	

210–200 mya

HERBIVOROUS, POSSIBLY OMNIVOROUS

50kg (110lb)

SOUTH WALES

TRIASSIC

RHAETIAN	
NORIAN	LATE
CARNIAN	
LADINIAN	MID
ANISIAN	
OLENEKIAN	EARLY
INDUAN	

(THEEK-oh-DONT-oh-sore-us)
THECODONTOSAURUS ANTIQUUS

200mya

HERBIVOROUS,
LOW BROWSER;
POSSIBLY
OMNIVOROUS

40kg
(90lb)

BRISTOL,
ENGLAND

The first identified Triassic dinosaur was once thought to have lived in an inland desert; now it is seen as an island-dweller, which could explain its small size. Around 200mya Britain sat just above the equator and along its western side lay the Mendip Archipelago, a scattering of tropical isles rich with plant and animal life. The islands formed from Carboniferous-era limestone. Sea levels were higher then; those islands are now the hilltops around Bristol and Avon, and many feature caves where the era's heavy rainfall and seawater dissolved pockets of limestone. The sea rose even higher at the end of the Triassic and filled the caves with sediments, which included a jumble of dinosaur bones. Over time these sediments turned into a different sort of limestone.

Skip forward 200m years to the 19th century, and humans began excavating the hillsides for building materials. In 1834 a crew of quarrymen found some mysterious bones, which they passed to Henry Riley, a doctor, and his naturalist friend Samuel Stutchbury. They examined a jaw embedded with 21 teeth and initially mistook it for an extinct lizard. The teeth were set in sockets, like mammals', rather than attached to the top of the jaw like most lizards', so they called it the 'socket-toothed reptile'. This turned out to be a vague name applicable to any dinosaur as all had teeth arranged that way – but they had no reason to know this as *Thecodontosaurus* was only the fourth dinosaur to be named. In 1975 another 11 specimens were found at Tytherington Quarry, north of Bristol, improving understanding of this primitive prosauropod's anatomy.

Then in 2008 came an unusual collaboration between Professor John Marshall, who studies pollen-grain fossils at the University of Southampton, and Dr David Whiteside, an expert in prehistoric reptiles at Bristol University. They analysed deposits within the cave sediments and detected fossilised pollen and algae that revealed the lush vegetation of the dinosaurs' habitat. Proving that it lived on the Mendip Archipelago rather than in an earlier Triassic desert, as previously supposed, raised the prospect that the 'Bristol Dinosaur' may have been a 'dwarf', as it was far smaller than similar near-contemporaries such as *Plateosaurus*. Living on an island can have this effect – see *Europasaurus* (page 115) and *Balaur* (page 309).

2.5m
(8ft)

RHAETIAN	LATE
NORIAN	
CARNIAN	
LADINIAN	MID
ANISIAN	
OLENEKIAN	EARLY
INDUAN	

220–205 mya

CARNIVOROUS

1100kg
(1 ton)

ARIZONA, TEXAS AND NEW MEXICO, USA

38

OTHER AMAZING ANIMALS OF THE TRIASSIC...

(LEP-toh-SOOK-us)

LEPTOSUCHUS CROSBIENSIS

12m
(39ft)

During the Late Triassic, giant, crocodile-like creatures inhabited the lakes and rivers, with genera such as *Leptosuchus* growing up to the length of a bus. Despite the resemblance they were not crocodiles but phytosaurs, a group of reptiles that developed a crocodilian appearance in a great example of convergent evolution – the process by which unrelated animals separately develop into a similar form by occupying a similar niche in an ecosystem.

Leptosuchus' fossils have been found in Texas, New Mexico and Arizona. While some of its relatives had long slender snouts that suggest a diet of fish, *Leptosuchus* had broad and powerful jaws that would have allowed it to attack terrestrial animals that came to the waterside to drink. However, the name phytosaur actually means 'plant-eating lizard', a misnomer derived from the fact that the first species identified in 1828 seemed to have a herbivore's teeth. These 'teeth' were actually lumps of fossilised mud! The phytosaurs were vicious carnivores and *Leptosuchus* was one of the mightiest examples.

(SHAS-ta-SORE-us)

SHASTASAURUS SIKKANNIENSIS

21m
(69ft)

TRIASSIC

RHAETIAN	LATE
NORIAN	
CARNIAN	
LADINIAN	MID
ANISIAN	
OLENEKIAN	EARLY
INDUAN	

210mya

PISCIVOROUS

68,000kg
(67 tons)

USA, CANADA
AND CHINA

This whale-sized ichthyosaur is the biggest marine reptile ever found. Its slender body and a snub, toothless snout made it quite different from most ichthyosaurs, which tended to have a long-nosed, dolphin-like form. The snout suggests that it was a suction feeder, subsisting on soft molluscs and fish. Ichthyosaurs were among the most important marine creatures of the Mesozoic era, existing in diverse shapes and sizes from the Triassic through to the Cretaceous. Features common to all species included enlarged eyes and a streamlined body. In 1902 *Shastasaurus* became the first large example documented after a fossil was found near Mount Shasta in California. Since then others have emerged in Canada and China.

39

TRIASSIC

RHAETIAN	
NORIAN	LATE
CARNIAN	
LADINIAN	MID
ANISIAN	
OLENEKIAN	EARLY
INDUAN	

(YOO-di-MORE-fo-don)

EUDIMORPHODON RANZII

This very early pterosaur skimmed the lakes of Late Triassic southern Europe hunting for fish. Experts know this because scales have been found in its fossils' stomach area and the long jaws bear more than 100 teeth of varying sizes – fangs at the front, small jagged multi-pointed teeth elsewhere – that equipped it well for gripping slippery specimens. Its claws allowed *Eudimorphodon* to clamber up cliffs and trees from where it could swoop down over the waters, and its long, stiff tail had a vane at the tip that may have helped control its aerial movement. Little is known about the animals from which it developed; all we know is that in the Late Triassic, this clearly proficient flyer pops up to provide evidence that avian reptiles were already well established, around 60m years before certain dinosaurs developed the ability for themselves. The fact that pterosaur bones were light enough to let them fly means they didn't fossilise as easily as other animals'. One possible ancestor of Triassic pterosaurs was *Heleosaurus*, a primitive, land-living animal that appears in the fossil record in the preceding Permian age, around 270mya. *Eudimorphodon*'s bones were first discovered in northern Italy and described in 1973.

Wingspan
1m (3ft)

210mya

P

PISCIVOROUS

10kg (22lb)

BERGAMO, ITALY

41

TRIASSIC

RHAETIAN	LATE
NORIAN	LATE
CARNIAN	LATE
LADINIAN	MID
ANISIAN	MID
OLENEKIAN	EARLY
INDUAN	EARLY

235–221 mya

H

HERBIVOROUS

80kg (175lb)

ARGENTINA AND BRAZIL

(EX-ee-REE-to-don)

EXAERETODON ARGENTINUS

Mammal-like herbivorous reptiles called cynodonts were prevalent throughout the Triassic. Towards the end of the age came specimens such as *Exaeretodon*, a squat, sturdy creature that seems to have resembled a dog but grew to the size of a large pig. This and several other species' fossils have been found in Carnian stage Triassic rock formations in Argentina and earlier Ladinian stage formations in Brazil. By the late Triassic the cynodonts were on the decline, their role as prime consumers of vegetation having been usurped by the rising sauropodomorph dinosaurs, but some kinds managed to survive all the way through to the Cretaceous.

Other species: *E. frenguelli, E. major, E. riograndensis, E. vincei*

1.8m (6ft)

(OL-ih-go-KIE-fus)
OLIGOKYPHUS
TRISERIALIS

TRIASSIC

RHAETIAN		LATE
NORIAN		
CARNIAN		
LADINIAN		MID
ANISIAN		
OLENEKIAN		EARLY
INDUAN		

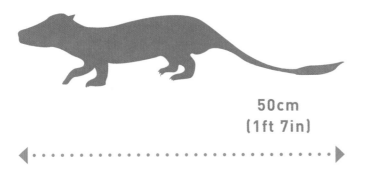

50cm
(1ft 7in)

227–180
mya

HERBIVOROUS

Weasel-like *Oligokyphus* species are known to have lived in Britain, Germany, North America and China from the Late Triassic through to the Early Jurassic. *Oligokyphus* was a synapsid, closer to the mammals than it was to other early synapsids such as the sail-backed *Dimetrodon*. Its strong teeth suggest a diet of seeds, nuts and tough plant material, and it may have been able to stand on its hind legs to nip leaves from bushes. *Oligokyphus* was named in 1922 and many specimens have been found since, confirming it as a very successful and prevalent little herbivore that scurried through woodlands alongside the increasingly dominant dinosaurs.

450g
(1lb)

Other species:
O. lufengensis,
O. major

BRITAIN,
GERMANY,
NORTH AMERICA
AND CHINA

QUIZ ON THE TRIASSIC

?

?

1. The Permian–Triassic extinction event, aka the 'Great Dying', that paved the way for the dinosaurs' emergence occurred how many million years ago?

2. Why was *Lessemsaurus*' back unusual compared with most Triassic dinosaurs?

3. Which is the earliest dinosaur known to have spent all its time on all fours?

4. What are the names of the two main orders of dinosaurs?

5. Which carnivore discovered in 2011 has a name meaning 'buck-toothed'?

6. *Pantydraco* takes its name from a village in which country?

7. What is the proper name of the Triassic herbivore known as 'the Bristol dinosaur'?

8. *Leptosuchus* looked like a giant crocodile but bore no relation to modern crocodilians. To which group of reptiles did it belong?

9. What term describes the process by which unrelated animals separately develop a similar appearance?

10. Which non-flying animal from 270mya in the Permian age is considered a possible ancestor of the pterosaurs?

For answers see page 340.

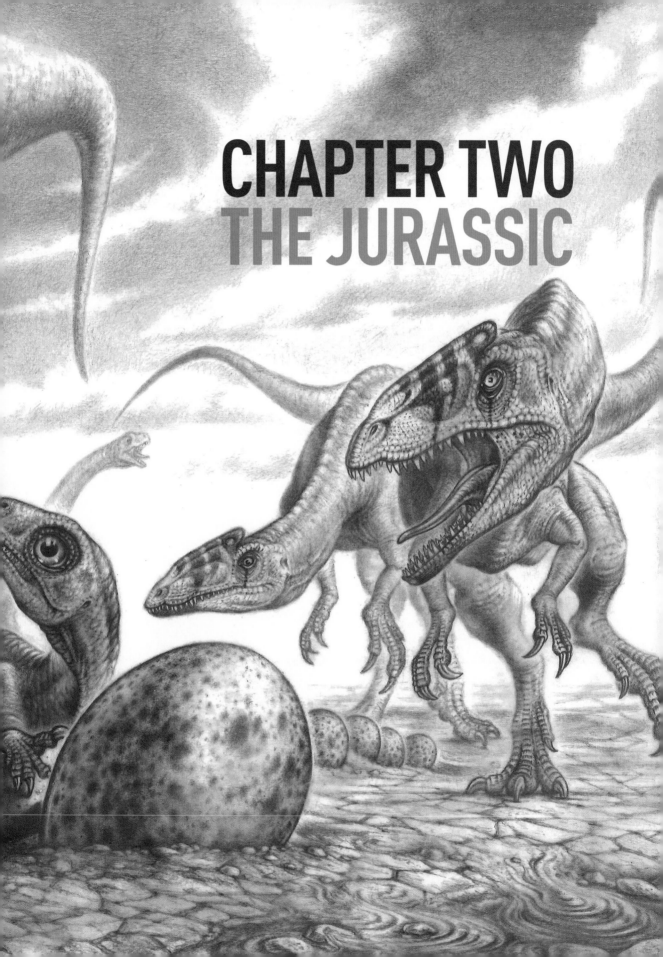

CHAPTER TWO
THE JURASSIC

THE JURASSIC

The Triassic–Jurassic extinction event saw the disappearance of more than half of the known species on land and in the ocean. Pangaea began to separate – the tectonic plates pulled apart and great ruptures and rifts formed that became inundated with seawater. This break-up of the landmasses sparked widespread volcanic activity that gradually killed great swathes of life on Earth.

But as the two new great landmasses, Laurasia to the north and Gondwana to the south, became surrounded by new seas, their inland temperatures dropped to habitable levels – and with this came a great blossoming of life. Pterosaurs soared through the skies and by the period's end they were joined by the first birds, while the ground rumbled throughout to the ever-heavier footsteps of huge herbivorous dinosaurs; the prosauropods of the Late Triassic developed now into the hugest animals to walk the Earth. Terrifying meat-eaters began to leave their three-toed prints across the land while even bigger carnivores ruled the seas. Terrifying pliosaurs began to rise, ichthyosaurs and plesiosaurs flourished, and the waters were rich with spiral-shelled ammonites, squid and fish.

With the moister, more temperate climate came lush vegetation, as former Triassic deserts grew green with ferns, cycads, horsetails and tall coniferous forests that helped satisfy the herbivorous dinosaurs' almighty appetite for foliage.

Named after the Jura Mountains on the Swiss-French border, where rocks from this period were first properly studied in the late 18th century, the Jurassic period saw the dinosaurs assume control of the Earth. It was a position they would not relinquish for 130m years.

JURASSIC

TITHONIAN	LATE
KIMMERIDGIAN	
OXFORDIAN	
CALLOVIAN	MIDDLE
BATHONIAN	
BAJOCIAN	
AALENIAN	
TOARCIAN	EARLY
PLIENSBACHIAN	
SINEMURIAN	
HETTANGIAN	

199–196 mya

HERBIVOROUS

500kg (1100lb)

SOUTH AFRICA

50

(ar-DON-ix)
AARDONYX CELESTAE

MEANING 'EARTH CLAW'

Worldwide interest greeted the announcement in 2009 of this bulky herbivore's discovery in South Africa, as it sits within an evolutionary gap that palaeontologists had been hoping to fill for years. We know the prosauropods of the Late Triassic and Early Jurassic, and the huge sauropods into which they eventually evolved by the Late Jurassic. *Aardonyx* fits neatly between the two in terms of its anatomy, posture and feeding habits.

It had the sauropodomorphs' elongated neck leading to a small head, the massive torso and long tail, but it was primarily bipedal. However, *Aardonyx* could drop down on to all fours and the shape of its forearm bones shows that they were in the process of evolving into front legs. They were beginning to interlock, making them stronger and less flexible; flexibility is not a good trait in limbs bearing a heavy load, as it would lead to the joints buckling and perhaps breaking. So here is the beginning of the skeletal structure that would allow sauropods such as *Argentinosaurus* to support a body weight of up to perhaps 80 tons.

Aardonyx also lacked the prosauropods' fleshy cheeks, which allowed it to open its mouth wider. This shows a transition between the prosauropods' way of picking at leaves and the sauropods' more industrial approach of 'bulk-browsing', in which they stripped branches of foliage wholesale.

Aardonyx's name combines Afrikaans and Greek and means 'earth claw', referring to the encrustation of haematite (a hard iron ore) around the fossil's toes, which made extracting the bones a very difficult task.

7m (23ft)

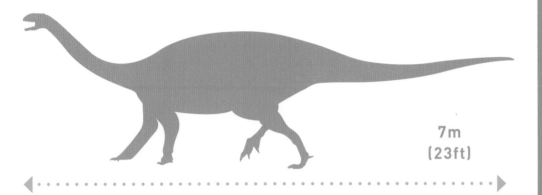

(yu-NAN-oh-SORE-us)

YUNNANOSAURUS HUANGI

7m
(23ft)

JURASSIC

TITHONIAN	
KIMMERIDGIAN	LATE
OXFORDIAN	
CALLOVIAN	
BATHONIAN	MIDDLE
BAJOCIAN	
AALENIAN	
TOARCIAN	
PLIENSBACHIAN	EARLY
SINEMURIAN	
HETTANGIAN	

200–185 mya

One of the latest prosauropods, *Yunnanosaurus* was especially notable for its teeth. Of the 20 or so skeletons found, two have skulls, and one of those has a set of 60 spoon-shaped teeth very like those of the later sauropods. The wear patterns show that, uniquely among prosauropods, *Yunnanosaurus* kept its teeth sharp by grinding them together when chewing foliage. However, the differences between the rest of its skeleton and the giant sauropods suggest that rather than being their direct ancestor, it evolved similar teeth to suit its environment and then they did so all over again tens of millions of years later. This is another example of convergent evolution, like the phytosaur *Leptosuchus* (page 38). The pioneering Chinese palaeontologist Yang Zhongjian (also known as C C Young) named *Yunnanosaurus* in 1942. When Lu Junchang described a second species in 2007, twice as large and dating from the Middle Jurassic, he named it in Young's honour.

Other species: *Y. youngi*

HERBIVOROUS, HIGH BROWSER

700kg (1500lb)

YUNNAN PROVINCE, SOUTHERN CHINA

51

JURASSIC

TITHONIAN
KIMMERIDGIAN
OXFORDIAN
LATE

CALLOVIAN
BATHONIAN
BAJOCIAN
AALENIAN
MIDDLE

TOARCIAN
PLIENSBACHIAN
SINEMURIAN
HETTANGIAN
EARLY

199–196
mya

H

HERBIVOROUS

uncertain

YUNNAN
PROVINCE,
SOUTHERN
CHINA

(YEE-zoo-SORE-us)
YIZHOUSAURUS SUNAE

This so-called 'missing link' sparked great excitement when revealed in 2010, though at the time of going to press it has not yet been officially described in a scientific paper. Understanding of the early sauropods' evolution into their immense descendants improved significantly when this exquisitely preserved specimen emerged from the Yunnan province of southern China. Although *Yizhousaurus* was far smaller at 10m (33ft) long, it had the big sauropods' quadruped stance, robust body and long neck – and most importantly the remains included a complete and perfect skull.

Sauropods' skulls were light and fragile, meaning they rarely fossilised, but here every detail was set in stone. It had a high, domed head with eyes on the sides, granting good awareness of approaching enemies. Its wide, U-shaped jaw was similar to the later *Camarasaurus*, according to Texas Tech University's

Professor Sankar Chatterjee, the palaeontologist who announced the find. Both jaws held sturdy serrated and spoon-shaped teeth that sliced up and down like scissors, cutting up vegetation as it fed. Fifty years before *Yizhousaurus'* discovery, the same rockbeds – known as the Lower Lufeng Formation – yielded remains of prosauropods such as *Lufengosaurus*. Thanks to *Yizhousaurus* we now have a better idea of how those primitive creatures evolved into the biggest animals ever to walk the Earth.

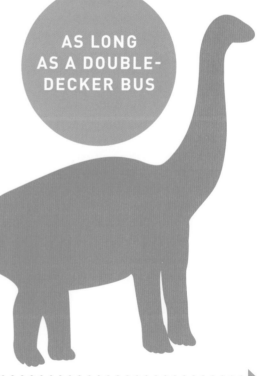

AS LONG
AS A DOUBLE-
DECKER BUS

10m
(33ft)

(SKOO-tell-oh-SORE-us)

SCUTELLOSAURUS LAWLERI

· ▶

1.3m
(4ft 3in)

◀ · ▶

JURASSIC

TITHONIAN	
KIMMERIDGIAN	LATE
OXFORDIAN	
CALLOVIAN	
BATHONIAN	MIDDLE
BAJOCIAN	
AALENIAN	
TOARCIAN	
PLIENSBACHIAN	EARLY
SINEMURIAN	
HETTANGIAN	

196mya

HERBIVOROUS

3kg
(7lb)

ARIZONA,
USA

The diminutive plant-eating *Scutellosaurus* was a predecessor to the great armoured beasts that would evolve later, such as stegosaurs and ankylosaurs. It had two ploys to defend itself from formidable predators such as the highly successful hunter *Coelyphysis*: escape by running as fast as it could, or hunker down and let its enemies try to penetrate its dense armour plating until they gave up. As a biped with strong legs and a long tail that helped it to balance, it would have been a quick mover compared with its later quadruped relatives. It is known from a small portion of skull and the bulk of two skeletons with loose armour plates, or 'scutes'. It had more than 300 of these small protective shields in six different forms, from bony lumps covering its back, to vertical plates like miniature versions of those later sported by *Stegosaurus*.

53

JURASSIC

TITHONIAN	
KIMMERIDGIAN	LATE
OXFORDIAN	
CALLOVIAN	
BATHONIAN	MIDDLE
BAJOCIAN	
AALENIAN	
TOARCIAN	
PLIENSBACHIAN	EARLY
SINEMURIAN	
HETTANGIAN	

195mya

C

CARNIVOROUS

**400kg
(900lb)**

**ARIZONA,
USA**

(di-LOWF-oh-SORE-us)
DILOPHOSAURUS WETHERILLI

Two semi-circular crests adorned the head of one of the first great carnivorous theropods. At 7m (23ft) long it is dwarfed by the monsters of the Late Cretaceous, but within the Early Jurassic *Dilophosaurus* would have been a domineering killer of large prosauropods and primitive armoured dinosaurs.

If you recall *Dilophosaurus* from the first *Jurassic Park* film, disregard your image of it. The real animal was far bigger and didn't spit poison – no dinosaurs are known to have been venomous, though some people argue that *Sinornithosaurus* (page 172) might have been.

Dilophosaurus' remains were discovered in northern Arizona in 1942 by a Navajo man named Jesse Williams. He informed a team of palaeontologists and they recovered three skeletons, none of which seemed to possess any skull ornamentation. Like many large theropods found in the early 20th century, the fossils were initially attributed to *Megalosaurus* (see page 64). But the discovery in 1964 of another specimen bearing a crest along the snout showed that this was a hitherto unknown animal. Re-examination of the best skull from the previous fossils showed a ridge where two such plates had broken away before fossilisation. In 1970 a new genus was created, with a name meaning 'double-crested lizard'.

**7m
(23ft)**

TITHONIAN	
KIMMERIDGIAN	LATE
OXFORDIAN	
CALLOVIAN	
BATHONIAN	MIDDLE
BAJOCIAN	
AALENIAN	
TOARCIAN	
PLIENSBACHIAN	EARLY
SINEMURIAN	
HETTANGIAN	

189–176
mya

OMNIVOROUS

220kg
(485lb)

NORTH-
EASTERN
USA

56

(AMM-oh-sore-us)
AMMOSAURUS MAJOR

MEANING
'SAND LIZARD'

When 19th-century construction workers built a new bridge in the Connecticut town of South Manchester, they included a hefty block of sandstone hewn from a local quarry. Later excavations at the quarry recovered the rear half of a skeleton of a medium-sized prosauropod but the rest was missing. It was only with the bridge's demolition in 1969 that the fossil was reunited with its front half and a full picture formed of this primitive omnivore. It would have been able to walk on two or four legs, and lay towards the smaller end of the sauropodomorph scale, which spans from the tiny *Saturnalia* of the Late Triassic to the 30m (98ft) *Argentinosaurus* that roamed through the Late Cretaceous. *Ammosaurus'* name means 'sand lizard', referring to the stone within which it was found.

up to 5m
(16ft)

(meg-APP-noh-sore-us)

MEGAPNOSAURUS RHODESIENSIS

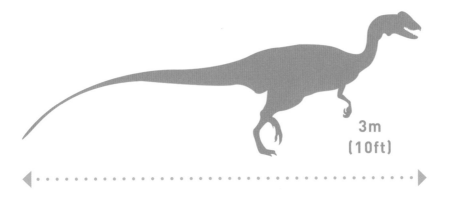

3m
(10ft)

JURASSIC

TITHONIAN	
KIMMERIDGIAN	LATE
OXFORDIAN	
CALLOVIAN	
BATHONIAN	MIDDLE
BAJOCIAN	
AALENIAN	
TOARCIAN	
PLIENSBACHIAN	EARLY
SINEMURIAN	
HETTANGIAN	

199–188 mya

C

CARNIVOROUS

32kg
(70lb)

ZIMBABWE AND SOUTH-WESTERN USA

Closely related to *Coelophysis*, this lean, agile, feathered theropod may have hunted in packs. Around 30 skeletons were discovered together in the 1960s in Zimbabwe (then known as Rhodesia). Later another species, *M. kayentakatae*, was discovered in Arizona in the south-western USA with small crests on its long, thin snout, prefiguring the larger ones sported by its later relative *Dilophosaurus*. Analysis of the remains suggested that *Megapnosaurus* had a lifespan of up to seven years.

Originally it was named *Syntarsus* by South African palaeontologist Mike Raath. That was in 1969; then more than 30 years later it emerged there was a beetle of the same name, so the dinosaur required renaming. The entomologists who spotted the error went ahead and did so, causing great controversy for several reasons. One, the name they chose was a joke: *Megapnosaurus* just means 'big dead lizard'. Two, it wasn't even accurate – this was a modestly-sized theropod for its time. Three, they were experts on insects rather than dinosaurs and were casually straying on to other scientists' territory. And four, by convention they should have consulted Raath. They didn't as they assumed he was dead – but he was still alive and left rather grumpy by this scientific *faux pas*. The rules dictate that once a new name is announced it must be used, so *Megapnosaurus* it is – unless scientists who believe it was actually a species of *Coelophysis* are vindicated, in which case it would take that name.

57

TITHONIAN	
KIMMERIDGIAN	LATE
OXFORDIAN	
CALLOVIAN	
BATHONIAN	MIDDLE
BAJOCIAN	
AALENIAN	
TOARCIAN	
PLIENSBACHIAN	EARLY
SINEMURIAN	
HETTANGIAN	

199–188 mya

H

HERBIVOROUS, HIGH BROWSER

135kg (300lb)

ZIMBABWE AND SOUTH-WESTERN USA

58

(MASS-oh-SPON-dih-lus)

MASSOSPONDYLUS CARINATUS

4m (13ft)

HIGH BROWSING

Thanks to the world's oldest known collection of dinosaur eggs, the way that prosauropods hatched their young is now coming to light after 190m years.

Gradual excavations since 2006 have revealed dozens of eggs laid by *Massospondylus* by a Jurassic lakeside, now preserved in stone in South Africa's Golden Gate Highlands National Park. They have been discovered in various layers, suggesting that dinosaurs used the site repeatedly over a long period of time. No nests were found, but the eggs were arranged in neat rows by the mothers after laying them. The embryos they contained were near-hatchlings of 15cm (6in) long. As these babies had no teeth and were clumsy walkers, it seems likely that their mothers cared for them for a period after birth. They also had four limbs of equal length, suggesting that they were quadrupedal, whereas the adults walked on two legs. They had longer necks than most prosauropods and a relatively slender build. Sir Richard Owen named *Massospondylus* in 1854, making it one of the earliest dinosaurs identified.

(See pages 46–47 for illustrations of *Massospondylus* and *Megapnosaurus*.)

(SAY-eet-AWD)

SEITAAD RUESSI

In Navajo folklore, Seit'aad is a monster that buries its victims in sand dunes, so the name seemed fitting for a little prosauropod whose white fossil bones were found entombed in the pale pink Navajo Sandstone rock formation in Utah. In the Early Jurassic the area was part of a vast desert. The finely preserved skeleton was found curled up as if it had been suddenly engulfed by sand; more typically dinosaurs are preserved in a 'death pose' with their neck arched backwards and head upright. The fossil's skull and tail had eroded away within the last millennium but the torso and limbs remained, showing Seitaad to be a rare North American example of an Early Jurassic prosauropod.

TITHONIAN		LATE
KIMMERIDGIAN		LATE
OXFORDIAN		
CALLOVIAN		
BATHONIAN		MIDDLE
BAJOCIAN		MIDDLE
AALENIAN		
TOARCIAN		
PLIENSBACHIAN		EARLY
SINEMURIAN		EARLY
HETTANGIAN		

185mya

H

HERBIVOROUS

90kg
(200lb)

4.5m
(15ft)

UTAH, USA

AS LONG AS
A GREAT
WHITE SHARK

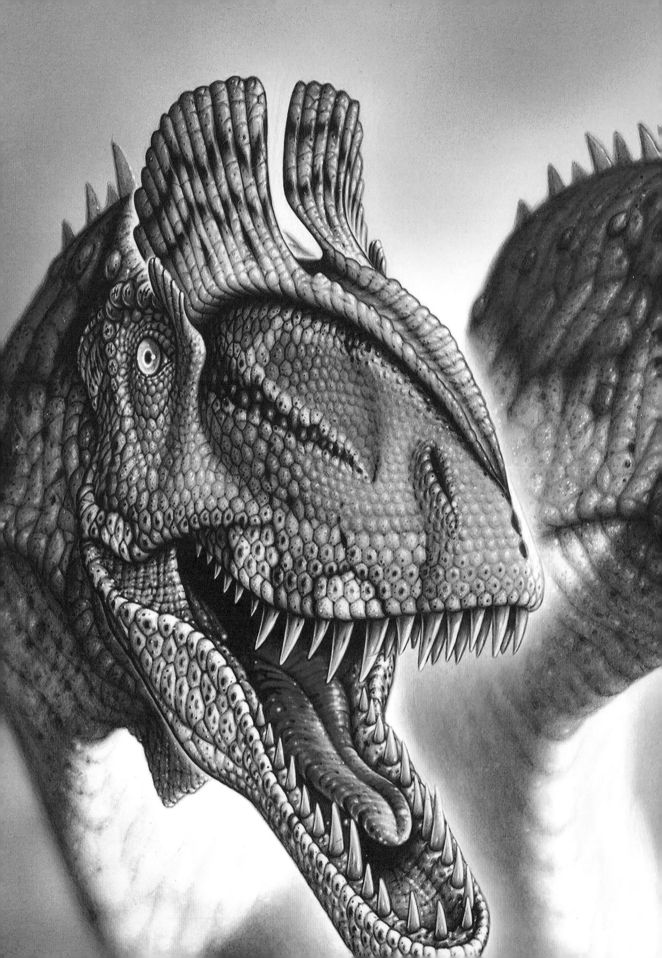

CRYOLOPHOSAURUS

The crested *Cryolophosaurus*, a major predator in Early Jurassic Gondwana, whose fossils were found on an Antarctic mountainside.

JURASSIC

TITHONIAN	
KIMMERIDGIAN	LATE
OXFORDIAN	
CALLOVIAN	
BATHONIAN	MIDDLE
BAJOCIAN	
AALENIAN	
TOARCIAN	
PLIENSBACHIAN	EARLY
SINEMURIAN	
HETTANGIAN	

190mya

C

CARNIVOROUS

465kg
(1025lb)

(CRY-oh-LOFF-oh-SORE-us)

CRYOLOPHOSAURUS ELLIOTI

6.5m
(21ft)

The so-called 'Elvisaurus' owes its nickname to a curious crest that ran across the top of its snout, resembling the legendary rock'n'roll star's pompadour hairstyle. It was probably related to *Dilophosaurus* (page 54), and in 1991 became the first carnivorous dinosaur discovered in Antarctica when its remains were recovered from 4000m (13,000ft) up a mountainside. Other discoveries at the site suggest that in the Early Jurassic this area was a forest also populated by a sauropodomorph

named *Glacialisaurus*, pterosaurs and mammal-like reptiles called tritylodonts, one of whose teeth was found in *Cryolophosaurus'* stomach. At the time the landmass was more than 600 miles further north and had a cool temperate climate. The single known specimen was 6.5m (21ft) long but was only a sub-adult so its full size is uncertain. It was one of the major predators in its environment but marks left on its bones show that it became a meal for other theropods after its death.

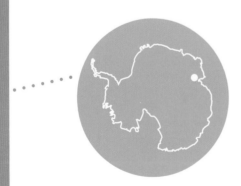

MOUNT
KIRKPATRICK,
ANTARCTICA

(KOH-ta-SORE-us)

KOTASAURUS YAMANPALLIENSIS

JURASSIC

TITHONIAN	
KIMMERIDGIAN	LATE
OXFORDIAN	
CALLOVIAN	
BATHONIAN	MIDDLE
BAJOCIAN	
AALENIAN	
TOARCIAN	
PLIENSBACHIAN	EARLY
SINEMURIAN	
HETTANGIAN	

Found in the Lower Jurassic stone of the Kota Formation in India, this is one of the earliest known sauropods, and like others it possessed the typical weighty body, long neck, long tail, plant-eating teeth and roughly horizontal posture. The hip and pelvic bones of 12 specimens found in the Andhra Pradesh state show a similarity to those of the prosauropods, the group's predecessors. The animals' remains were found jumbled in a former riverbed, and it is thought that a herd drowned in a flood and were washed to this point.

196–183 mya

HERBIVOROUS

ONE OF THE EARLIEST KNOWN SAUROPODS

2500kg (2.4 tons)

9m (30ft)

SOUTH-EASTERN INDIA

63

JURASSIC

TITHONIAN	
KIMMERIDGIAN	LATE
OXFORDIAN	
CALLOVIAN	
BATHONIAN	MIDDLE
BAJOCIAN	
AALENIAN	
TOARCIAN	
PLIENSBACHIAN	EARLY
SINEMURIAN	
HETTANGIAN	

166mya

CARNIVOROUS

2000kg
(1.9 tons)

**ENGLAND
AND FRANCE**

64

(MEG-ah-lo-SORE-us)
MEGALOSAURUS BUCKLANDII

The first dinosaur ever identified was a devastating killer that roamed Britain in the middle of the Jurassic period – but the animal itself is often overlooked in favour of its significance to the story of dinosaur discovery.

Megalosaurus is, if you like, the original dinosaur: the history of our knowledge of them begins here. That history opened in 1815 in a quarry near Oxford. A seam of limestone known as 'Stonesfield slate' had long borne mystifying fossil remains, and now a few more bones emerged including a huge jawbone embedded with teeth. In 1824 the Very Rev William Buckland, Oxford University's first Professor of Geology, described them in a study titled 'Notice on the *Megalosaurus* or great fossil lizard of Stonesfield'. In 1842 Richard Owen proclaimed this animal a member of the Dinosauria alongside *Iguanodon* and *Hylaeosaurus*. These strange creatures, albeit inaccurately understood, soon caught the public imagination; by 1853 *Megalosaurus* had even appeared at the atmospheric beginning of Charles Dickens' novel *Bleak House*. The great author conjures a dismal November day: 'As much mud in the streets as if the waters had but newly retired from the face of the Earth, and it would not be wonderful to meet a *Megalosaurus*, forty feet long or so, waddling like an elephantine lizard up Holborn Hill.'

Of course we know now that it didn't waddle, it wasn't quite 12m (40ft) long, and it was far from elephantine. It was a muscular killer that probably hunted stegosaurs and sauropods, bounding at some speed before clawing its prey with powerful arms and biting with blade-like serrated teeth set into a hefty, long-jawed skull. It may also have scavenged carcasses. When it was discovered, *Megalosaurus* seemed huge but it was actually an average-sized theropod, with reasonably large arms that probably bore three fingers. It had four toes, three of them touching the ground, leaving an imprint that fits well with a series of menacing-looking footprints striding across exposed stone at Ardley, Oxfordshire: compelling evidence, were it needed, that terrible creatures once roamed Great Britain, and *Megalosaurus* was as horrifying as any of them.

9m
(30ft)

WASTEBASKET GENUS

A 'wastebasket genus' is the term given to a classification of animal to which numerous others are mistakenly assigned without sufficient study. Owing to its position as the first known dinosaur, *Megalosaurus* was the first dinosaur to assume this status: around 50 other Middle Jurassic theropods were at times placed in the *Megalosaurus* genus before being granted names in their own right, among them *Carcharodontosaurus*, *Dryptosaurus* and *Proceratosaurus*. If you imagine discarding all the knowledge we now have of the dinosaur era's incredible diversity, it's easy to understand why early researchers might assume that all these superficially similar theropods were the same animal.

(MAG-no-SORE-us)

MAGNOSAURUS NETHERCOMBENSIS

JURASSIC

TITHONIAN	LATE
KIMMERIDGIAN	
OXFORDIAN	
CALLOVIAN	MIDDLE
BATHONIAN	
BAJOCIAN	
AALENIAN	
TOARCIAN	EARLY
PLIENSBACHIAN	
SINEMURIAN	
HETTANGIAN	

MEANING 'GREAT LIZARD'

Among the many carnivores that have spent a period being mistaken for *Megalosaurus* is this fellow European theropod, a pacey predator whose remains were found in Nethercombe, Dorset, in the 19th century. (If a species name ends in *-ensis*, this means the first part is the location where it was found.) In 1932 Friedrich von Huene reclassified it as *Magnosaurus*, reflecting their similarity by coining a name with the same meaning: great lizard. However, like many inhabitants of the *Megalosaurus* 'wastebasket genus' that are known from fragmentary remains, it received little further study for many years. Then in 2010 the young Cambridge University palaeontologist Roger Benson took a detailed look at *Megalosaurus* and its many associated species and concluded that *Magnosaurus* is definitely a different animal, primarily as its jawbone has a combination of characteristics unseen in any other dinosaur. He positioned it as one of the earliest known members of the Tetanurae, or stiff-tailed theropods. That makes *Magnosaurus* significant as the ancestor of better known dinosaurs as varied as *Tyrannosaurus*, *Archaeopteryx* and *Velociraptor*.

175mya

CARNIVOROUS

500kg
(1100lb)

4m
(13ft)

DORSET,
ENGLAND

JURASSIC

TITHONIAN	
KIMMERIDGIAN	LATE
OXFORDIAN	
CALLOVIAN	
BATHONIAN	MIDDLE
BAJOCIAN	
AALENIAN	
TOARCIAN	
PLIENSBACHIAN	EARLY
SINEMURIAN	
HETTANGIAN	

170mya

C

CARNIVOROUS

45kg
(100lb)

WESTERN
AUSTRALIA

(OZ-rap-tor)

OZRAPTOR SUBOTAII ··············· AND ··

2m
(6ft 6in)

Australians seem fond of allotting dinosaurs names that leave no doubt about their country of origin. There's *Australovenator* (page 226), *Austrosaurus* and the more informal *Ozraptor*, which is only known from an 8cm-long (3in) fragment of shinbone but is likely one of the earliest abelisaurs – muscular, near-horizontal speedy predators with high, short skulls and minuscule arms that reached their apex in the southern hemisphere during the Cretaceous. It also seems probable that *Ozraptor* is Australia's joint-oldest known dinosaur alongside a sauropod called *Rhoetosaurus*.

Four schoolboys found the bone protruding from rocks at Bringo Railway Cutting, near the city of Geraldton, in the mid-1960s. They alerted the University of Western Australia, whose experts removed the slab of sandstone containing the fossil and sent a cast to the Natural History Museum in London. The reply came that it was probably from a Jurassic turtle, but in the 1990s palaeontologists gained a clearer idea after finally removing the bone from its matrix. It proved to be part of a theropod's leg with an unusual ankle joint that seems adapted so it could run especially fast. From that scrap of bone found by a group of boys, palaeontologists can now picture a predator that prowled the woodlands of Jurassic Australia.

In the far larger *Eoabelisaurus* we have another contender for the earliest known abelisaur – after these two, the next recorded lived 40m years later. This is reflected in a name meaning 'dawn abelisaur' (remember *Eoraptor* and *Eodromaeus* (pages

JURASSIC

TITHONIAN	
KIMMERIDGIAN	LATE
OXFORDIAN	
CALLOVIAN	
BATHONIAN	MIDDLE
BAJOCIAN	
AALENIAN	
TOARCIAN	
PLIENSBACHIAN	EARLY
SINEMURIAN	
HETTANGIAN	

(EE-oh-ay-BEL-ih-SORE-us)

EOABELISAURUS MEFI

22–23), the 'dawn plunderer' and 'dawn runner' from back in the Triassic). The typical form was beginning to develop: the skull was not as markedly short as in, say, *Carnotaurus* (page 312) nor the arms as withered, but the beginnings of those modifications are discernible. While its arms were of a normal length, the hand is tiny. It seems therefore that the diminishing of abelisaurs' forelimbs happened in stages: first the hands shrunk, then the arms much later.

When *Eoabelisaurus* was alive, Gondwana had not split into northern and southern land-masses. In naming it in 2012,

Argentinian palaeontologist Diego Pol and his German colleague Oliver Rauhut therefore noted that its near-complete skeleton's discovery in Patagonia prompts the question of why no abelisaurs have been found in rocks the same age in northern Europe. The conclusion is that even when Gondwana was connected some unsurpassable obstacle lay between the north and south. In their paper describing *Eoabelisaurus*, Pol and Rauhut note 'growing evidence from climate modelling and geological data for a large, central Gondwanan desert during the Middle and Late Jurassic' – that is, a desert so vast that dinosaurs could not cross it.

174–168 mya

CARNIVOROUS

900kg (1980lb)

CHUBUT, PATAGONIA, ARGENTINA

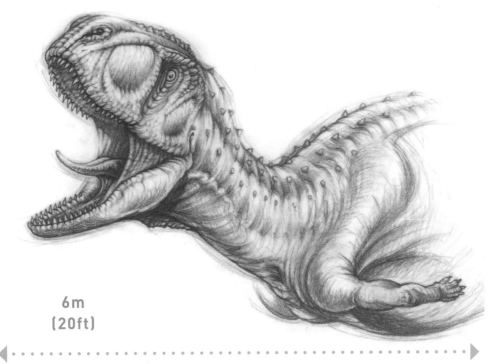

6m
(20ft)

JURASSIC

TITHONIAN	
KIMMERIDGIAN	LATE
OXFORDIAN	
CALLOVIAN	
BATHONIAN	MIDDLE
BAJOCIAN	
AALENIAN	
TOARCIAN	
PLIENSBACHIAN	EARLY
SINEMURIAN	
HETTANGIAN	

165–161 mya

H

HERBIVOROUS, LOW BROWSER

3000kg (2.9 tons)

CENTRAL CHINA

70

(SHOO-no-SORE-us)
SHUNOSAURUS LII

AND

Spiky tails were thought to be the preserve of stegosaurs – and then *Shunosaurus* came along. Its bones were discovered in China in 1977 but not studied closely until more than a decade later, when researchers revealed that it bore a cluster of conical osteoderms at the tip of its tail that it swang at would-be predators.

But it seemed unique among sauropods until *Spinophorosaurus'* discovery in Niger in 2009 provided further spectacular evidence of this behaviour. *Spinophorosaurus* was a bigger dinosaur with an even more dangerous tail. The huge skeleton found coiled in red siltstone in the Niger desert suggested that its tail was not clubbed but bore two symmetrical pairs of long, sharp spikes almost identical to those wielded by stegosaurs.

LOW BROWSING

9.5m (31ft)

(SPY-no-FOR-oh-SORE-us)

SPINOPHOROSAURUS NIGERENSIS

13m (43ft)

Shunosaurus was one of the earliest eusauropods, the advanced group of 'true sauropods' whose characteristics include a tiny head, elongated neck and massive size. But being a very primitive example, few of these characteristics are well developed in *Shunosaurus*; for instance it was very small by later eusauropod standards and actually had a very short neck, the second shortest known after that of *Brachytrachelopan*. Like that animal, *Shunosaurus* was perhaps a low browsing feeder.

JURASSIC

TITHONIAN		
KIMMERIDGIAN		LATE
OXFORDIAN		
CALLOVIAN		
BATHONIAN		MIDDLE
BAJOCIAN		
AALENIAN		
TOARCIAN		
PLIENSBACHIAN		EARLY
SINEMURIAN		
HETTANGIAN		

170–165 mya

HERBIVOROUS

7000kg (6.8 tons)

NIGER, AFRICA

71

VICTORIANS

'And God created the great sea-monsters, and every living creature that creepeth, wherewith the waters swarmed, after its kind, and every winged fowl after its kind; and God saw that it was good.'

Genesis 1.21

That was on the fourth day of the Creation, according to the Bible. In the early 19th century most British people read Genesis as a literal truth and believed that this six-day creation occurred a few thousand years ago. Numerous scholars had already devoted much effort to calculating the exact year, the best known being the 17th-century Irish Archbishop of Armagh, James Ussher: by his reckoning, God created the world on the night before Sunday, 23rd October, 4004BC.

But as the 19th century progressed, so did the distance between the foundations of orthodox religion and the hard evidence emerging from England's quarries and cliffs, in tandem with the calculations of geologists and palaeontologists. In his *Principles of Geology*, published from 1830 to 1833, Charles Lyell observed that the Earth's rock formations had accumulated, and its landmasses very gradually drifted apart, over a tremendous expanse of time. Imagine being sure that the world was 5000 years old and then being confronted with evidence that ultimately would prove it was almost a million times older! (The Earth is now reckoned to be 4.6bn years old, and the universe 13bn years old.)

At the same time scientists were developing early theories of evolution,

arguing that way back into the 'deep time' proposed by Lyell's studies, there existed some common ancestry between humans and diverse other animals.

For many people these radical ideas were immensely hard to accept, a seismic shift that moved the ground beneath their feet. This drastic broadening of the vista of life on Earth threw society into a dizzy spin, disorienting anyone who clung to the old beliefs – which often included the people who found and scientifically studied the fossils. The growing realisation that ancient fragments of bone and teeth that littered Britain's rock formations were relics of a lost world of terrible lizards only further unsettled some of Georgian and Victorian society's most fundamental convictions. In 1824 William Buckland described the dinosaur that he named *Megalosaurus*, whose sharp teeth were indisputably shaped for killing animals and consuming their flesh – but this did not fit with a literal reading of the Bible, which states that death and the eating of meat only developed after the Fall of Man.

As the man who brought *Megalosaurus* to public prominence, Buckland embodied the conflict between scientific progress and religious orthodoxy at the dawn of dinosaur research. His full title was the

eyesight. Buckland was just as prone to bizarre behaviour: he was a zoophage, someone who enjoys eating animals of every description. Dishes on the menu at his Deanery included battered mouse, squirrel pie, roast panther, porpoise, mole and bluebottle flies, the last two being the only creatures he concluded were unfit for human consumption. This habit actually tallied with his religious belief in humanity's divinely conferred position as Earth's apex predator, for whom God had placed all other animals as sources of nutrition. His academic style was unorthodox: once during a lecture he thrust a hyena's skull in an undergraduate's face and demanded: 'What rules the world?' When no answer was forthcoming from the scared, baffled student, Buckland declared: 'The stomach, sir, rules the world. The great ones eat the less, the less the lesser still!'

In his *Bridgewater Treatise* he argued that successive waves of creation had paved the way for humanity's arrival on Earth but even he later had to buckle in the face of mounting evidence for gradual change across millions of years.

The first half of the century's developments culminated in 1859 when Charles Darwin, until then best known as

Very Rev Dr William Buckland, for as well as being an Oxford University geologist he also became in 1845 the Dean of Westminster. In 1820 he published a book called *Vindiciæ Geologiæ; or the Connexion of Geology with Religion explained*, in which he sought to fit the emerging sense of 'deep time' and evidence of prehistoric life into a revised understanding of divine creation. In 1836 he returned to the subject in one of the eight *Bridgewater Treatises* funded by a bequest from the late Francis Egerton, the Earl of Bridgewater, who was a noted eccentric; he gave dinner parties for dogs which he clothed in the fashions of the time, and kept pigeons in his garden with clipped wings so that he could shoot them despite his poor

the geologist aboard *HMS Beagle* and much influenced by Lyell's ideas, published *On the Origin of Species* to enormous controversy. The slow transformative processes that Lyell applied to geology, Darwin transferred to biology. His work built upon previous evolutionary theories and provided compelling evidence in support, ensuring that within 20 years his theories were broadly accepted by scientists and the public alike. The initial effect, however, was to polarise opinion and antagonise the religiously devout (and while his ideas are the basis of scientific orthodoxy today, they still have this effect on many people). A famous cartoon of the time depicts Darwin with a monkey's body, ridiculing his proposal that humans and apes share a distant ancestor rather than having been created in a fixed form. *Archaeopteryx*'s discovery in 1861 seemed serendipitous, a perfect demonstration of his ideas about transitional species, in this case showing how dinosaurs began to evolve into bird-like forms.

But even among some of his peers Darwin's ideas proved controversial. Lyell himself, being a devout Christian, had difficulty accepting Darwin's theory of evolution by natural selection rather than divine creation. Sir Richard Owen was one of Victorian Britain's most eminent anatomists but neither could he accept Darwin's ideas about the 'transmutation' of one species into another; he just considered *Archaeopteryx* a bird, and preferred an earlier evolutionary theory proposing that life evolved within fixed 'archetypes'. Humans could change over time and so could gorillas, but Owen couldn't countenance a connection between the two.

Owen could be a difficult character who had so little respect for his peers that he was eventually removed from the Royal Society's council for failing to credit other scientists' work in his publications. His intense focus on his work extended to his domestic life; as the era's pre-eminent zoologist he had first refusal on the carcasses of animals from London Zoo, and one day his wife returned home to find a dead rhino in the hallway. But he was an important scientist whose memorable achievements included establishing the Natural History Museum and coining the word 'dinosaur', from the Greek for 'terrible lizard', in 1842 to describe the aforementioned *Megalosaurus*, along with *Iguanodon* and *Hylaeosaurus*. His conception of dinosaurs as squat, lumpen reptiles is made concrete in the Crystal Palace sculptures of numerous prehistoric creatures including those three, which were built by Benjamin Waterhouse Hawkins to Owen's specifications and stand today as a relic of early dinosaur palaeontology.

After Darwin returned from his voyage on the *Beagle*, Lyell introduced him to Owen and they were friendly for a time. Then their relationship soured horribly in the aftermath of *On the Origin of Species*' publication. The mild-mannered Darwin noted in 1860 that 'the Londoners say

he is mad with envy because my book is so talked about. It is painful to be hated in the intense degree with which Owen hates me'. And after Owen attempted in 1871 to cut the government's funding to Kew Gardens' botanical collection, Darwin reflected: 'I used to be ashamed of hating him so much, but now I will carefully cherish my hatred and contempt to the last days of my life.' This indicates the emotions coursing through scientific circles in the high Victorian era, and the significance that these people rightly attached to their work.

However, while Owen was a tricky character we should not scoff at his misconceptions about dinosaurs' appearance, nor at his and other 19th-century scientists' efforts to reconcile religion with the evidence emerging from rockbeds across Europe and America. Buckland, Mantell, Owen, Lyell and Darwin pushed back the boundaries of contemporary thought and thus laid the first rungs on the palaeontological ladder; without them modern experts could not have climbed to the elevated view they enjoy today.

Besides, we are hardly at the top of that ladder and have much further to ascend. Given the rate at which new evidence is emerging, in another century's time some of our present assumptions about the dinosaurs' evolutionary relationships, appearance and behaviour are likely to look very dated indeed.

JURASSIC

TITHONIAN	
KIMMERIDGIAN	LATE
OXFORDIAN	
CALLOVIAN	
BATHONIAN	MIDDLE
BAJOCIAN	
AALENIAN	
TOARCIAN	
PLIENSBACHIAN	EARLY
SINEMURIAN	
HETTANGIAN	

164–161 mya

H

HERBIVOROUS

16,000kg
(15.7 tons)

NORTHERN
AFRICA

(jo-BAR-ee-ah)
JOBARIA TIGUIDENSIS

A 19-person team led by American palaeontologist Paul Sereno found almost the entire skeleton of this primitive sauropod in Niger's Sahara Desert in 1997. They were led to the site by local Tuareg tribespeople, who knew of the old bones there and associated them with a mythical animal they called Jobar. It turned out that the remains at this mass graveyard were actually those of a previously unknown dinosaur. One juvenile specimen's bones bore bite-marks fitting the teeth of *Afrovenator*, the 'African Hunter' that was a prime carnivore in Jurassic northern Africa.

Jobaria may have been a macronarian, which were giant herbivores distinguished by their long necks and large snouts (the term literally means 'big nose'). The sediments its bones lay in were initially ascribed to the Early Cretaceous but a study in 2009 suggested them to be far older. At the time this area of northern Africa was covered by woodland. *Jobaria* is distinguished by its relatively short tail – at least, when compared with the slender, whip-like tails of the diplodocids – and its unusually simple vertebrae. Tests carried out by Sereno and his team later showed that its hind legs were strong enough to allow it to rear up and eat high leaves beyond smaller herbivores' reach.

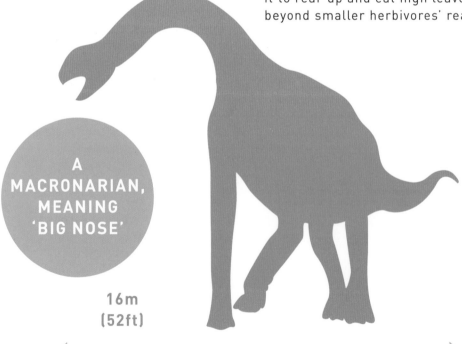

A MACRONARIAN, MEANING 'BIG NOSE'

16m
(52ft)

(gwaan-LONG)
GUANLONG WUCAII

JURASSIC

TITHONIAN	
KIMMERIDGIAN	LATE
OXFORDIAN	
CALLOVIAN	
BATHONIAN	MIDDLE
BAJOCIAN	
AALENIAN	
TOARCIAN	
PLIENSBACHIAN	EARLY
SINEMURIAN	
HETTANGIAN	

160mya

3.5m
(11ft 6in)

C

CARNIVOROUS

125kg
(275lb)

The strange crest on its snout makes *Guanlong* appear most unusual, but its scraping teeth, large skull and powerful legs connect it with a very famous group of dinosaurs. The line of aggressive carnivores known as tyrannosaurs reached its apex in *Tyrannosaurus rex*, and 90m years earlier this is one its first known ancestors. *Guanlong* is known from two adult and juvenile skeletons found in a remote corner of northern China; the species name means 'five colours', and refers to the local rock formations.

Its later relative *Dilong* was feathered so *Guanlong* may also have been, though no feather impressions were preserved. It had three long, powerful fingers, of far more use than the late tyrannosaurs' tiny digits. Its name, meaning 'crowned dragon', refers to its crest, which may have been brightly coloured and was probably for display, being too thin and weakly constructed to serve any other purpose. Announcing the find, Dr James Clark of George Washington University noted this ornamentation's resemblance to those found on modern hornbill and cassowary birds.

JUNGGAR BASIN, NORTH-WEST CHINA

77

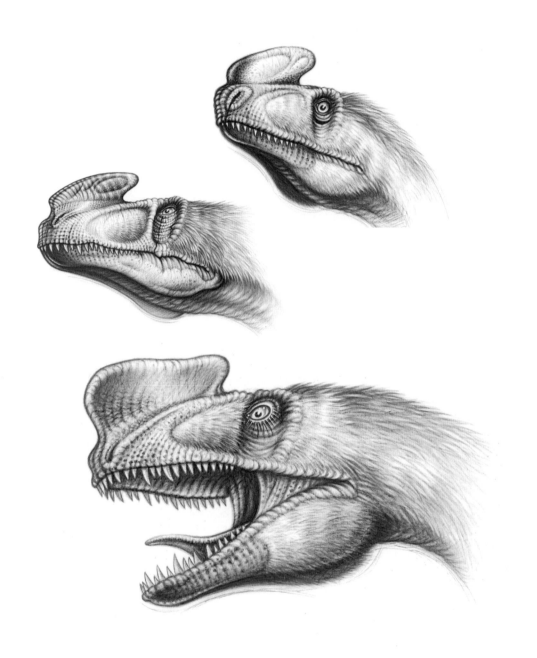

3m
(10ft)

(PRO-seh-RAT-oh-SORE-us)

PROCERATOSAURUS BRADLEYI

The elongated jaw of this small English theropod was unearthed in 1910 but it took another century for *Proceratosaurus* to be confirmed as another very early tyrannosauroid.

At only 30cm long its skull was a fifth the length of *T. rex*'s, but the two have much in common: the same kind of gaps on either side for increasing the jaw muscles, the same banana-shaped teeth and, within the bone, numerous air pockets that lighten the structure.

The fossil was discovered during excavations for a reservoir and presented in 1942 to the Natural History Museum, where it remains. It had a small crest on its snout, which prompted German palaeontologist Friedrich von Huene's view that it was an ancestor of the Late Jurassic dragon-like predator *Ceratosaurus*, hence the name. But that dinosaur was a relative of the abelisaurs, whereas a study in 2010 showed *Proceratosaurus* to be a tyrannosauroid, thus making its name misleading.

It gives its name to the Proceratosauridae family, of which other members include *Sinotyrannus* and *Kileskus*, pictured opposite, above left and above right.

JURASSIC

TITHONIAN	
KIMMERIDGIAN	LATE
OXFORDIAN	
CALLOVIAN	
BATHONIAN	MIDDLE
BAJOCIAN	
AALENIAN	
TOARCIAN	
PLIENSBACHIAN	EARLY
SINEMURIAN	
HETTANGIAN	

165mya

C
CARNIVOROUS

50kg
(110lb)

MINCHINHAMPTON, GLOUCESTERSHIRE, ENGLAND

79

JURASSIC

TITHONIAN	
KIMMERIDGIAN	LATE
OXFORDIAN	
CALLOVIAN	
BATHONIAN	MIDDLE
BAJOCIAN	
AALENIAN	
TOARCIAN	
PLIENSBACHIAN	EARLY
SINEMURIAN	
HETTANGIAN	

170–165
mya

H

HERBIVOROUS,
LOW BROWSER

300kg
(660lb)

SANGONGHE
VALLEY,
CHINA

(TYAN-chee-SORE-us)

TIANCHISAURUS NEDEGOAPEFERIMA

MEANING
'HEAVENLY
LAKE LIZARD'

Ankylosaurs are most often associated with the Cretaceous but this one, the earliest example known, dates back to the Middle Jurassic. It had a small flattened club at the end of its tail and an armoured back; its descendants would develop both features to extremes, making them almost invulnerable to attack.

Tianchisaurus briefly had a different name that not only referred to its era but to the film that prompted its study. This armoured herbivore's bones were collected in 1974 by geology students on a fieldwork trip but it was only in 1993 that it was described and named, thanks to funding from *Jurassic Park* director Steven Spielberg. Because of this it was initially known as *Jurassosaurus nedegoapeferima*, the clumsy species name referring to the surnames of his film's cast: Sam Neill, Laura Dern, Jeff Goldblum, Richard Attenborough, Bob Peck, Martin Ferrero, Ariana Richards and Joseph Mazzello. Chinese palaeontologist Dong Zhiming eventually opted for a title meaning 'Heavenly Lake lizard' (a reference to the famous location in the Tianshan Mountains where it was discovered) but he retained the species name.

3m
(10ft)

(DAT-oo-SORE-us)
DATOUSAURUS BASHANENSIS

JURASSIC

TITHONIAN	
KIMMERIDGIAN	LATE
OXFORDIAN	
CALLOVIAN	
BATHONIAN	MIDDLE
BAJOCIAN	
AALENIAN	
TOARCIAN	
PLIENSBACHIAN	EARLY
SINEMURIAN	
HETTANGIAN	

180mya

10m
(33ft)

AS LONG AS
TWO CARS

HERBIVOROUS

The Dashanpu Quarry in China is the world's most prolific source of dinosaur bones from the Middle Jurassic. Among the 40 tonnes of bones that it has revealed in recent years are the remains of this herbivore, which is notable for being one of only a few sauropods known from circa 180mya; most emerge in the fossil record at the end of the Jurassic and into the Cretaceous.

Datousaurus had a long neck compared with its contemporaries but had nothing to compare with extraordinary later Chinese sauropods such as *Mamenchisaurus*. Its name means 'chieftain lizard', and it is known from two partial skeletons, one of which came with a partial skull that suggests an unusually heavy head for a sauropod, whose skulls were typically so light that they rarely remained intact long enough to fossilise.

4500kg
(4.4 tons)

SICHUAN
PROVINCE,
CHINA

81

JURASSIC

TITHONIAN	
KIMMERIDGIAN	LATE
OXFORDIAN	
CALLOVIAN	
BATHONIAN	MIDDLE
BAJOCIAN	
AALENIAN	
TOARCIAN	
PLIENSBACHIAN	EARLY
SINEMURIAN	
HETTANGIAN	

168–164
mya

C

CARNIVOROUS

700kg
(1500lb)

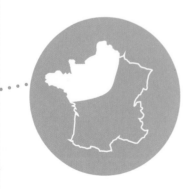

**NORTH-WEST
FRANCE**

82

(du-BROY-oh-SORE-us)

DUBREUILLOSAURUS VALESDUNENSIS

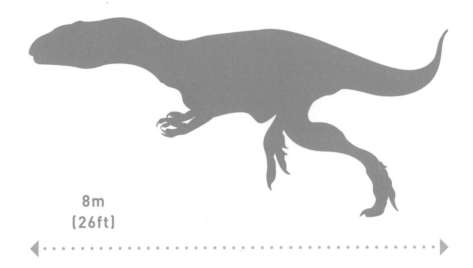

**8m
(26ft)**

Dubreuillosaurus's most distinctive feature was its elongated head, which was three times as long as it was high. When this strange skull and a few ribs were uncovered at a disused French quarry in 1998 it was initially considered a new species of the far bigger *Poekilopleuron* (see opposite). By the time researchers could return to tease out further bones, the quarry had reopened and the bulldozers had succeeded in smashing the rest of the fossil remains into 2000 tiny fragments that lay scattered around the site. After several years' painstaking work collecting and analysing these shards of ancient bone, French palaeontologist Ronan Allain announced in 2005 that this was a hitherto unknown Middle Jurassic megalosaur, which he named after the family who made the original discovery. As such it has, for non-French speakers at least, one of the hardest dinosaur names to pronounce! This bulky, stooping carnivore would have hunted its prey in the coastal mangroves of northwestern Laurasia.

(PO-kil-oh-PLURE-on)

POEKILOPLEURON BUCKLANDII

· · · · · · · · ·▶

Powerful, muscular arms set this theropod apart: while certain later big carnivores' forelimbs would diminish to the point of apparent uselessness, for this mighty predator living 100m years earlier in the Middle Jurassic they were evidently a major part of its hunting technique. *Poekilopleuron* was a bulky beast akin to *Megalosaurus* and relatively slow on its feet, so it probably grappled with ungainly herbivores such as stegosaurs and sauropods. Its name is a badly constructed blend of Greek and Latin meaning 'varied ribs', because very unusually the fossil contained a complete ribcage, which comprised three distinct types of bone.

9m
(30ft)

JURASSIC

TITHONIAN	
KIMMERIDGIAN	LATE
OXFORDIAN	
CALLOVIAN	
BATHONIAN	MIDDLE
BAJOCIAN	
AALENIAN	
TOARCIAN	
PLIENSBACHIAN	EARLY
SINEMURIAN	
HETTANGIAN	

170–165 mya

CARNIVOROUS

1000kg
(0.9 tons)

NORMANDY, FRANCE

83

TITHONIAN	
KIMMERIDGIAN	LATE
OXFORDIAN	
CALLOVIAN	
BATHONIAN	MIDDLE
BAJOCIAN	
AALENIAN	
TOARCIAN	
PLIENSBACHIAN	EARLY
SINEMURIAN	
HETTANGIAN	

164–161
mya

HERBIVOROUS,
LOW BROWSER

15kg
(33lb)

SICHUAN
PROVINCE,
CHINA

84

(a-JILL-oh-sore-us)

AGILISAURUS LOUDERBACKI

During the construction of the Zigong Dinosaur Museum in China, workers uncovered one of the most complete skeletons known of a small ornithischian dinosaur. As its name suggests, this was a very agile and graceful little creature, which could evade attack by running at great speeds on its long hind legs. When walking or feeding it probably stood on four legs with its short forelimbs inclining its small head towards the ground, so that it could easily graze on plants such as ferns.

The *Agilisaurus* remains were found at the richly fossiliferous Dashanpu Quarry. The Zigong museum on the site displays a cast of that *Agilisaurus* skeleton beside the frightening frame of a *Gasosaurus*, which would have been one of its chief predators. *Gasosaurus* was a theropod whose curious name refers to its fossil's discovery during the construction of a gasworks in China's Sichuan province.

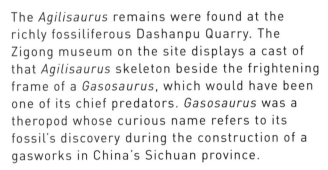

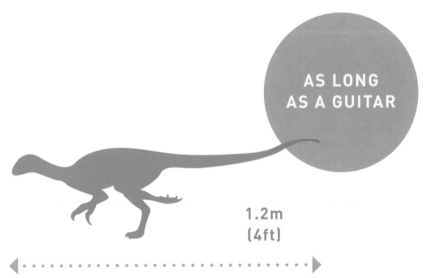

AS LONG
AS A GUITAR

1.2m
(4ft)

(SEE-tee-oh-sore-ISK-us)

CETIOSAURISCUS STEWARTI

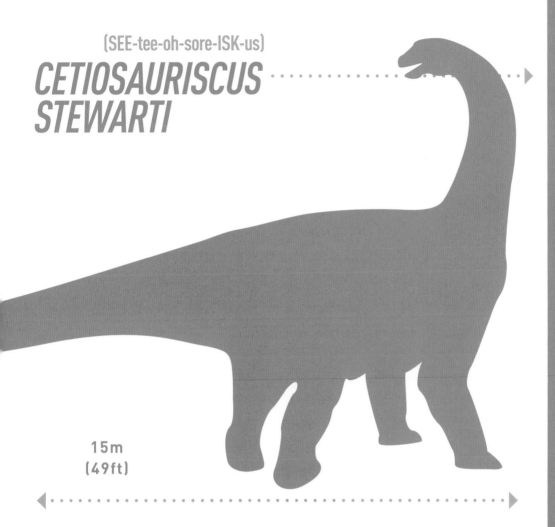

JURASSIC

TITHONIAN	
KIMMERIDGIAN	LATE
OXFORDIAN	
CALLOVIAN	
BATHONIAN	MIDDLE
BAJOCIAN	
AALENIAN	
TOARCIAN	
PLIENSBACHIAN	EARLY
SINEMURIAN	
HETTANGIAN	

163mya

15m
(49ft)

HERBIVOROUS

4000kg
(3.9 tons)

This herbivore, the height of a typical terraced house, ambled around the woodlands of Middle Jurassic England, browsing leaves and now and again trying to fend off attacks by carnivores such as *Megalosaurus*. Its partial remains, a series of vertebrae and a front leg, are hard to classify – different studies have placed it as a diplodocoid and a mamenchisaurid. Friedrich von Huene named it in 1927 when he reclassified some fossils that had previously been attributed to *Cetiosaurus*, the 'whale-like lizard'. This sauropod was so similar that he granted a new name meaning 'like the whale-like lizard'. The species name honours the chairman of the London Brick Company, whose Oxford Clay limestone quarry at Peterborough yielded the first *Cetiosauriscus* fossils.

PETERBOROUGH AND CIRENCESTER, ENGLAND

85

TITHONIAN

KIMMERIDGIAN

OXFORDIAN

CALLOVIAN

BATHONIAN

BAJOCIAN

AALENIAN

TOARCIAN

PLIENSBACHIAN

SINEMURIAN

HETTANGIAN

LATE

MIDDLE

EARLY

160mya

C

CARNIVOROUS –
LIZARDS, SMALL
MAMMALS

25kg
(55lb)

NORTH-
WESTERN
CHINA

(HAP-lo-KEER-us)

HAPLOCHEIRUS SOLLERS

MEANING 'SIMPLE, SKILFUL HAND'

This feathered, sharp-toothed carnivore looks distinctively bird-like – but it lived around 15m years before *Archaeopteryx*, which is often considered the first known bird. *Haplocheirus* was the largest and most primitive known alvarezsauroid. Note the last three letters of that word: this means it was a close relative of the alvarezsaurids but did not actually lie within the Alvarezsauridae family. It lived 63m years earlier than any known alvarezsaurids, taking their roots back to long before the birds' emergence. This confirmed alvarezsaurids had a common ancestry with birds but were not birds themselves, merely a strange offshoot that evolved birdlike characteristics in parallel.

Because alvarezsaurids differ in many ways from modern birds, their removal from the dinosaur-to-bird evolutionary line reduced a confusing element and thus strengthened the connection.

Later members of the family, such as *Mononykus* (page 287) had short and extremely muscular arms with a single huge thumb-claw that was probably used for breaking into termite mounds. Here the impressive claw is just beginning to evolve from a more typically theropod grasping hand, which has three strong digits with the middle one particularly long. Its name means 'simple, skilful hand', referring to the fact it could have used its fingers for catching prey in a way that alvarezsaurids could not. Scientists thought that the alvarezsaurids of the Cretaceous must have had predecessors in the Jurassic, and *Haplocheirus*' discovery proved them right.

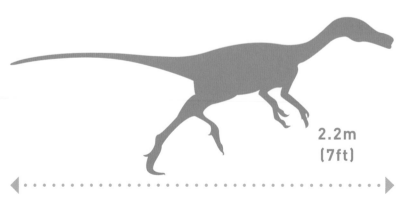

2.2m
(7ft)

(CON-dor-rap-tor)

CONDORRAPTOR CURRUMILI

JURASSIC

TITHONIAN	
KIMMERIDGIAN	LATE
OXFORDIAN	
CALLOVIAN	
BATHONIAN	MIDDLE
BAJOCIAN	
AALENIAN	
TOARCIAN	
PLIENSBACHIAN	EARLY
SINEMURIAN	
HETTANGIAN	

Argentinian farmer Hippolito Currumil found a single tibia bone on his land in the village of Cerro Condor, and in 2005 German palaeontologist Oliver Rauhut revealed it to be from a new genus of small predatory dinosaur. But it remained poorly understood until 2007, when the world's first Middle Jurassic theropod skeleton with intact joints came to light in South America. It revealed *Condorraptor* to be a fast-running killer, and in 2010 Cambridge palaeontologist Roger Benson confirmed its more precise status as a megalosauroid closely related to its South American contemporary *Piatnitzkysaurus*.

As Middle Jurassic South American theropod fossils are few and far between, further study may help fill a gap in knowledge of the better known later carnivores' evolution.

164mya

C

CARNIVOROUS

200kg (440lb)

SOUTHERN ARGENTINA

4.5m (15ft)

87

THE BONE WARS

America, the mid-19th century. Pioneers and prospectors spread westward across the land, laying claim to this 'new' world (though of course it wasn't new to the Native Americans who had lived there since time immemorial). Within this context of discovery and competition, bloodshed and intrigue, the 'Gilded Age' that made the modern USA, let us picture two eminent Americans.

One was a dandyish character, high-born into a wealthy Quaker family. He sported short but luxuriant hair, a glint in his eyes, a fulsome moustache, a defiant jutting jaw and the hint of a smirk on his lips. His name was Edward Drinker Cope.

The other, Othniel Charles Marsh, was a sturdy, jowly fellow whose long, dark and wispy beard distracted from the lack of hair on his head. His immediate family were of more modest means; but his uncle was George Peabody, one of the 19th century's foremost industrialists. It was Peabody's generosity that enabled Marsh to begin an academic career that took him to Germany for a spell, Europe then being at the frontier of fossil research, and back to America in 1866 to become Professor of Vertebrate Palaeontology at Yale University. That year also saw the 35-year-old Marsh

convince his uncle to establish the Peabody Museum of Natural History, which opened at the Yale campus in Connecticut with Marsh installed as its director. Three years later Peabody died, leaving Marsh a huge inheritance that formed the foundation of his future endeavours. By this time Cope was a professor of zoology and a member of the Academy of Natural Sciences, a venerable institution in Philadelphia.

The two men had met in Berlin in 1864. For a time they were colleagues, even friends. Then, as each strove for success, a rivalry arose that turned into a full-blown feud. By the height of their careers they were sworn enemies, but out of their attempts to outdo one another came a proliferation of discoveries that brought the age of the dinosaurs to public prominence and shifted palaeontology's focus from Europe to the USA.

The Bone Wars began in earnest when Cope and his team were working the marl pit in New Jersey where in 1858 Joseph Leidy, the greatest of the previous generation of American fossil-hunters, helped unearth the first *Hadrosaurus* skeleton. Marsh heard about the site's rich potential and bribed Cope's team of bone-diggers to send their finds to him rather than to Cope. When Cope learned of this he was livid.

Another incident around this time consolidated their mutual loathing. In 1868 Cope received and reconstructed a fossil of a giant plesiosaur, which he named *Elasmosaurus* in a scientific journal. Marsh took immense pleasure in publicly announcing that Cope's published description of the fossil had the skull placed at the wrong end of the spine. A humiliated Cope attempted to buy and destroy every remaining copy of the journal in question, and from here there was no going back: the coming 20 years saw their enmity escalate to ever-greater heights. In 1877 both men received a sample of fossils from Arthur Lakes, a Colorado geologist and artist who had found some huge bones while out hiking. Not knowing that Lakes had also contacted Cope, Marsh paid him $100 to keep the discovery a secret – and then on learning the truth, hurriedly sent an employee to Colorado to secure ownership of the find. By now their rivalry was growing famous, as was Marsh's willingness to dispense large sums of money from the comfort of his desk at Yale; he was less inclined than Cope to get his hands dirty at the rockface,

Edward Drinker Cope,
1840–1897

Othniel Charles Marsh,
1831–1899

although he understood the value of travelling to meet Native Americans who knew the badlands far better than most white men. In the 1870s Marsh befriended a Sioux chief named Red Cloud and received a tribal nickname: Wicasa Pahi Hohu, or Man That Picks Up Bones. They also called him Big Bone Chief.

When a pair of workers laying the Union Pacific Railroad ploughed into a rich seam of Late Jurassic fossils at Como Bluff in Wyoming, they notified Marsh of the ancient bones they had turned up, hinting that if he did not make a suitably generous financial offer then they would alert Cope instead. Sure enough, Marsh paid out. The investment proved worthwhile: in the December 1877 edition of the *American Journal of Science* Marsh unveiled newly discovered dinosaurs including *Stegosaurus*, *Allosaurus* and *Diplodocus*.

The following years would bring accusations of spying, theft and bribery. In a demonstration of how the pair's initial ideals concerning scientific progress were overwhelmed by their all-consuming egos, they even took to destroying fossils to prevent them from falling into one another's hands. They would disguise their own excavation sites by filling them with dirt and rocks and at one stage a stone-throwing fight broke out when their two teams were working in close proximity.

Como Bluff was the site of a showdown in 1879 when Cope arrived and accused Marsh of trespassing and stealing his fossils. Marsh subsequently ordered that the pits his team were excavating should be dynamited, rather than run the risk of Cope finding any fossils there.

For his part, Cope had a train packed with Marsh's finds diverted to Philadelphia. Marsh tried to hinder and confuse Cope by littering his sites with fragments of bones from other places. And on it went, the two men exhausting their spirits and their finances as they tried to outdo one another.

With 80 discoveries to Cope's 64, Marsh was the victor in the Bone Wars, though neither man emerged well from the saga. Their behaviour blemished their reputations and their fierce competition led at times to a lack of scientific exactitude that saw many of their discoveries reclassified by later experts. Cope died aged 56 in 1897, almost bankrupted by the cost of his extravagant expeditions. Marsh died two years later aged 67, also in financial difficulty.

Palaeontology was the real winner: the two men's achievements combined to revolutionise our understanding of the dinosaurs' era and bring a lost world into the modern public eye.

JURASSIC

TITHONIAN	
KIMMERIDGIAN	LATE
OXFORDIAN	
CALLOVIAN	
BATHONIAN	MIDDLE
BAJOCIAN	
AALENIAN	
TOARCIAN	
PLIENSBACHIAN	EARLY
SINEMURIAN	
HETTANGIAN	

165–153 mya

INSECTIVOROUS

2kg (5lb)

LIAONING, CHINA

(zhow-TIN-gee-ah)
XIAOTINGIA ZHENGI

This long-tailed, chicken-sized feathered dinosaur hunted insects in the woodlands of Jurassic China... but the most important thing to know about it is the effect its discovery had on our perception of a far more famous creature that lived at least 5m years later: *Archaeopteryx*. *Xiaotingia*'s unveiling in 2011 prompted headlines claiming that *Archaeopteryx* had been 'knocked off its perch' as the first bird, but the truth is more nuanced.

A prolific Chinese palaeontologist Xu Xing described *Xiaotingia* as a very close relative of *Archaeopteryx*, arguing that the pair were not birds and should both be classified as deinonychosaurs, the group of dinosaurs that also includes *Troodon* and *Velociraptor*. At the time he backed this up by citing a large hole at the front end of the snout, close to the nostril. This is a feature only seen in deinonychosaurs and not present in birds.

Xu also noted that *Archaeopteryx* and *Xiaotingia* show early signs of developing one of the deinonychosaurs' definitive features, an extendable second toe.

So if *Archaeopteryx* were a deinonychosaur and not a bird, what was the first bird? Contenders include three other little feathered theropods found in China: *Epidexipteryx*, *Jeholornis* and *Sapeornis*. However, the notion of 'the first bird' really isn't very scientifically helpful, as it suggests a simplicity that doesn't help us understand the gradual process of evolution.

In any case, some months after *Xiaotingia* was announced, another analysis classified it as a troodontid and moved *Archaeopteryx* back into the birds again. The authors of this paper, Michael Lee and Trevor Worthy, argued that Xu's classification was weakly supported by evidence but noted: 'Both positions for *Archaeopteryx* remain plausible, highlighting the hazy boundary between birds and advanced theropods.'

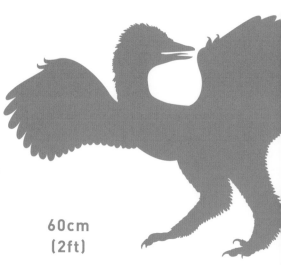

60cm (2ft)

BIRDS OF A FEATHER?

This is one of the most confusing and contested areas in palaeontology today. Bear in mind that while all birds are dinosaurs, not all dinosaurs are birds. Creatures such as *Archaeopteryx* and *Xiaotingia* were certainly dinosaurs and certainly had bird-like qualities; the big question is whether they sit somewhere on the branch of the tree of life that leads to the birds we know today.

After all, many of the attributes that led to *Archaeopterx*'s definition as a bird – its feathers, wishbone and three-fingered hand, for example – have since been found in non-avian dinosaurs. *Tyrannosaurus* had two of those three characteristics, and it may well have had feathers too, but no one would call it a bird or a direct ancestor of modern birds. It's just that at some stage far further back, *T. rex* and the sparrow share a common ancestor from which they inherited those traits.

The fact that these feathered theropods could probably fly or glide does not define them as birds either: bats can fly, so can insects, and way back in time so could pterodactyls. It may only mean that a branch of the dinosaur family that lies outside the birds' lineage also evolved flight.

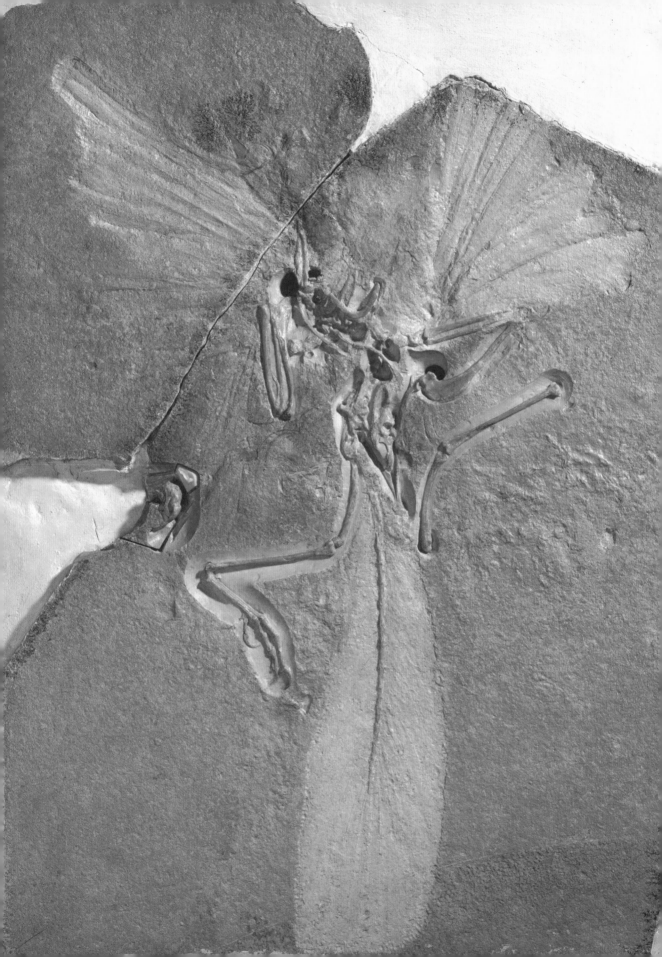

(ar-kee-OP-ter-ix)
ARCHAEOPTERYX LITHOGRAPHICA · · · · ▶

We all know about *Archaeopteryx* – it is an icon of evolution, a creature whose discovery in Germany's Solnhofen beds in 1861 resonated with the recent publication of *On the Origin of Species* and seemed to vindicate Charles Darwin's hypothesis about the gradual development of life. Germans called it the '*Urvogel*', the 'first bird'. Here was a creature with the jaws and tail of a dinosaur but, unmistakeably preserved in 147m-year-old stone, delicate feathers fanning from its limbs and tail. The beauty of the fossil reinforced the creature's renown. But it is 150 years since *Archaeopteryx*'s discovery and while it was revelatory then, similar creatures now emerge from Chinese rocks with thrilling regularity. Nowadays we understand it as just one of many small, feathered theropods that flapped and fluttered through the Jurassic woodlands.

Its fossil is very familiar – but what do we know of the living *Archaeopteryx*? When this magpie-sized bird (or not, as the case may be) existed, that part of Europe consisted of tropical islands in a warm sea, on a similar latitude to Florida today. In all likelihood it could fly, or at least glide from the trees, as it had an arrangement of feathers suiting flight protruding from its arms. Its hind legs were also covered in 'trousers' of feathery down, and feathers fanned from the tail, but its fossils do not show any feathers on the head or neck. It may be that *Archaeopteryx* was bald-headed like some modern birds, or that they just weren't preserved in the fossils. It probably ate insects, small animals and perhaps fish, and may have been able to swim. An extended claw on its foot helped it catch larger prey and defend itself, as did its hand claws.

50cm
(1ft 7in)

JURASSIC

TITHONIAN	
KIMMERIDGIAN	LATE
OXFORDIAN	
CALLOVIAN	
BATHONIAN	MIDDLE
BAJOCIAN	
AALENIAN	
TOARCIAN	
PLIENSBACHIAN	EARLY
SINEMURIAN	
HETTANGIAN	

150–147 mya

INSECTS, SMALL ANIMALS, FISH?

1kg (2.2lb)

GERMANY

95

DINOSAUR BIRD
CONNECTIONS

When you compare illustrations of feathered and beaked dinosaurs with the birds that we see around us today, the similarities are so striking that it is a wonder anyone disputes the notion of a clear evolutionary link between the two.

For a small band of palaeontologists the issue is far from resolved: some argue instead that birds descended from a Triassic reptile named *Longisquama*, which had a row of feathers – or similar upright bristly structures – protruding from its spine. The mainstream view, however, is to divide dinosaurs into two kinds: non-avian and avian. The non-avians are the long-

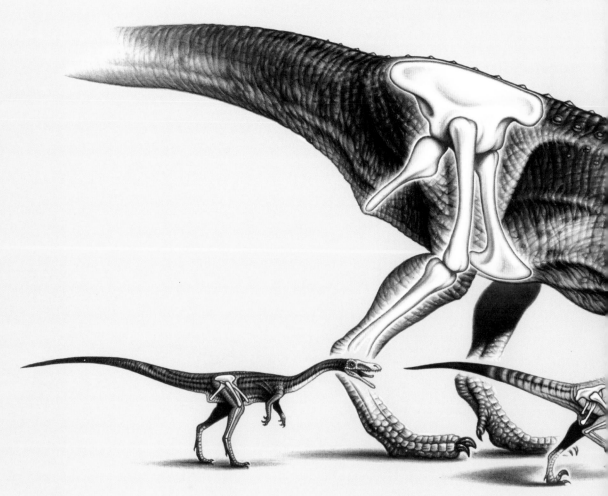

extinct reptiles that we traditionally think of as dinosaurs. The avian ones span from ancient flying creatures such as *Archaeopteryx* (page 95) right through to the sparrows twittering in the hedgerow and, more obviously, large flightless birds such as the ostrich and emu. Yes, the birds we know today are dinosaurs – which may initially seem amazing but makes perfect sense when you compare their characteristics. The consensus is that of the two great groups of dinosaurs, bird-hipped and lizard-hipped, modern birds actually evolved from the latter: in particular from the carnivores known as theropods, and specifically from the small feathered hunters known as maniraptorans.

Today, palaeontologists can present reams of similarities between theropods and birds, from their joints' formation and the microscopic structure of their eggs, to their feathers and the way their lungs work. Take *Velociraptor* as an example. Like modern birds, it brooded over a nest of eggs and had a wishbone, light hollow bones, soft internal organs and scales such as birds have on their feet. Palaeontologists proved in 2007 that *Velociraptor* had feathers by analysing a specimen's ulna, which along with the radius is one of the two bones that make up a forearm – this is the same for any vertebrate, whether a dinosaur or a human being. They found faint impressions of 'quill knobs', the slight indentations that show where feathers' quills were implanted into the bone.

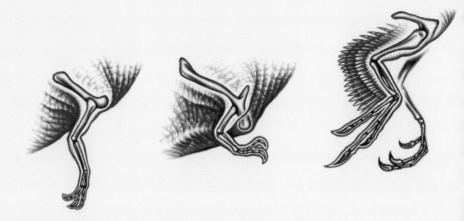

But the most astonishing work that might just strengthen the dinosaur-to-bird connection took place the same year under the guidance of American palaeontologist Mary Higby Schweitzer. Her team claimed to have discovered blood vessels and blood cells in a 68-million-year-old *Tyrannosaurus* legbone, and on examination these tissues revealed a structural similarity to those of birds such as ostriches and emus. However, this research has stirred great controversy and other scientists believe that the tissues could result from later contamination rather than belong to the *Tyrannosaurus*. Meanwhile in a related study John Asara and colleagues at Harvard Medical School claimed to have obtained protein sequences from collagen recovered from the same *Tyrannosaurus* bone. Of all the animals alive today, *T. rex*'s amino acid sequences turned out to bear the closest resemblance to those of... the chicken.

Schweitzer and her colleagues later reported that they had found medullary bone from the hind leg of a *Tyrannosaurus*. Medullary bone was previously thought unique to female birds, which create it to keep their limb bones strong when they are diverting calcium into growing eggs. Its discovery formed another link between dinosaurs' and birds' physical make-up.

The idea of their being related stretches back to the mid-19th century when dinosaurs were first identified. Thomas Henry Huxley, nicknamed 'Darwin's Bulldog' for his aggressive advocacy of his friend's evolutionary theories, hailed *Archaeopteryx* as a missing link between dinosaurs and birds. More generally he observed distinct similarities between birds' skeletons and those of theropod dinosaurs. In September 1876 he gave a long lecture in New York in which he noted the blurred boundaries, stating: 'We have had to stretch the definition of the class of birds so as to include birds with teeth and birds with paw-like fore limbs and long tails. There is no evidence that *Compsognathus* possessed feathers; but, if it did, it would be hard indeed to say

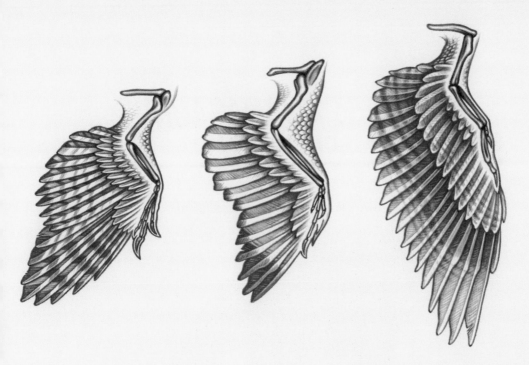

whether it should be called a reptilian bird or an avian reptile.'

But in 1926 Gerhard Heilmann published *The Origin of Birds*, which dismissed the connection, helping to usher in a four-decade 'quiet era' in palaeontology during which it was assumed that dinosaurs were a dull evolutionary dead-end.

A century after *Archaeopteryx*'s discovery, however, John Ostrom (1928–2005) described a find that would reignite the debate and confirm him as one of the foremost palaeontologists of the 20th century.

Working in Montana in 1964, Ostrom found fossil remains of a medium-sized maniraptoran, a close relation of *Velociraptor*, that he would name *Deinonychus* (page 213). Ostrom's study of *Deinonychus* revealed that its forearms were very similar to *Archaeopteryx*'s. Ostrom also found numerous features held by meat-eating theropods such as *Deinonychus* that are only present in birds among modern life-forms.

The lineage is very complicated. Over hundreds of millions of years there have been parallel evolutions in different parts of the world, and recurrences of traits among primitive birds that other extinct animals had developed aeons earlier. It seems some birds developed the ability to fly, then died out, and then others evolved the ability again separately much later. And, as ever, new evidence could change everything: for instance if a fossilised bird were discovered dating from before the dinosaurs that are generally considered the birds' ancestors.

But that is most unlikely – what seems far more plausible to most experts in the early 21st century is that the dinosaurs did not die out altogether. They are here among us today, pecking seeds from the bird-feeder in your back garden, dabbling and diving on the pond in your local park, and sitting on your plate at Sunday lunch. Next time you carve a roast chicken and save the wishbone, remember that you're seeing part of the evidence that connects it with a *Tyrannosaurus rex*.

JURASSIC

TITHONIAN	
KIMMERIDGIAN	LATE
OXFORDIAN	
CALLOVIAN	
BATHONIAN	MIDDLE
BAJOCIAN	
AALENIAN	
TOARCIAN	
PLIENSBACHIAN	EARLY
SINEMURIAN	
HETTANGIAN	

155–145 mya

HERBIVOROUS

23,000kg
(22.6 tons)

UTAH AND COLORADO, USA

(CAM-a-ra-SORE-us)

CAMARASAURUS SUPREMUS

**Other species:
C. grandis,
C. lentus,
*C. lewisi***

18m
(59ft)

Camarasaurus was the most common herbivore on the plains of the Late Jurassic western USA. Groups of fossils have been found that suggest it roamed in herds. Its blunt, box-like head contained teeth stronger than most sauropods', which suggests that it could eat tougher vegetation than others, chewing it up rather than gulping it down wholesale. *Camarasaurus* fossils show no evidence of gastroliths – stones swallowed and held in the gut to aid digestion, common in other herbivores – which reinforces this theory.

'Camara' comes from the Greek for 'chamber', just as in the device you use to take photographs. Here it refers to the chambers that Edward Drinker Cope found in the dinosaur's vertebrae. They reduced the weight of *Camarasaurus*' frame: a feature later found to be common in sauropods, but less well documented in 1877 when Cope described the first sauropod found in America.

JURASSIC

TITHONIAN	
KIMMERIDGIAN	LATE
OXFORDIAN	
CALLOVIAN	
BATHONIAN	MIDDLE
BAJOCIAN	
AALENIAN	
TOARCIAN	
PLIENSBACHIAN	EARLY
SINEMURIAN	
HETTANGIAN	

(AL-oh-SORE-us)

ALLOSAURUS FRAGILIS

This most famous of Jurassic carnivores was abundant across the American Midwest where it ruled as top predator, employing its serrated knife-like teeth, 25cm-long (10in) ripping claws and powerful arms to kill sauropods, many of whose bones have been found bearing chips and scrapes from *Allosaurus* teeth. Its huge hind legs and muscular S-shaped neck provided enormous attacking strength, and once its victim was dead, its jaws could widen enabling it to swallow huge lumps of flesh. In short, this was a model of the perfected voracious hunter – no wonder it rose to become one of the great successes of the Jurassic period.

Individuals of all ages are known; the remains of 44 *Allosaurus* specimens have been found at Cleveland-Lloyd Quarry in Utah alone. Othniel Marsh coined its name in 1877.

155–150 mya

CARNIVOROUS

1700kg (1.6 tons)

8.5m (28ft)

COLORADO AND UTAH, USA

JURASSIC

TITHONIAN	
KIMMERIDGIAN	LATE
OXFORDIAN	
CALLOVIAN	
BATHONIAN	MIDDLE
BAJOCIAN	
AALENIAN	
TOARCIAN	
PLIENSBACHIAN	EARLY
SINEMURIAN	
HETTANGIAN	

160–145
mya

H

HERBIVOROUS

36,000kg
(35.4 tons)

SZECHUAN,
CHINA

(ma-MEN-chi-SORE-us)

MAMENCHISAURUS HOCHUANENSIS

The extraordinary *Mamenchisaurus* had a neck containing 19 vertebrae and comprising half of its entire body length. This well-documented species' neck probably measured 9.5m (31ft); another species, *M. sinocanadorum*, is known from a poorer fossil but its neck may have been 11m (36ft) long. So if they used their apparent great height to eat leaves from the highest treetops in the forests of prehistoric China, how did a *Mamenchisaurus*' heart pump enough blood all the way up to its brain for it to remain conscious while doing so?

One answer is that it didn't adopt such a giraffe-like posture at all, and actually raised its neck no more than 20 degrees above horizontal. Which prompts the next question: if the purpose wasn't to become immensely tall and eat leaves beyond others' reach, why did it evolve such a long neck? Some experts suggest that actually *Mamenchisaurus* – and perhaps other especially long-necked sauropods – operated more like a vacuum cleaner, moving its head in a slow and methodical sweep from side to side, scouring foliage across a broad distance without having to waste any energy moving its feet. However, others argue that such sauropods evolved solutions to such problems that have been posed about blood pressure, and that their long necks really were used to reach up to great height and feed from treetops. *Mamenchisaurus* is among China's best-known dinosaurs.

25m (82ft)

JURASSIC

TITHONIAN	
KIMMERIDGIAN	LATE
OXFORDIAN	
CALLOVIAN	
BATHONIAN	MIDDLE
BAJOCIAN	
AALENIAN	
TOARCIAN	
PLIENSBACHIAN	EARLY
SINEMURIAN	
HETTANGIAN	

160mya

C

CARNIVOROUS

900kg
(1980lb)

ENGLAND

(MET-ree-ah-CANTH-oh-SORE-us

METRIACANTHOSAURUS PARKERI

MEANING 'MODERATELY SPINED LIZARD'

This medium-sized carnivore hunted sauropods in what is now England. In the 1800s its fragmentary remains were found at Jordan's Cliff, near Weymouth, by the geologist and fossil collector James Parker. *Altispinax* had a high ridge along its spine; *Metriacanthosaurus* probably had a smaller one, hence a name meaning 'moderately spined lizard'.

It was a close relative of the better understood Chinese carnivore *Yangchuanosaurus*, and is likewise possibly a sinraptorid. However, that group's most distinguishing features were its skulls, and the single known *Metriacanthosaurus* fossil didn't include one. Judging from those few heightened vertebrae, sections of hip, thighbone and shinbone, its exact relations are difficult to determine.

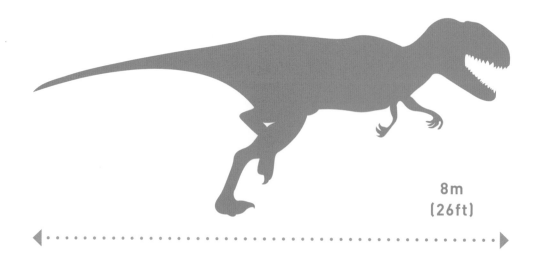

8m
(26ft)

(TOO-oh-JANG-oh-SORE-us)
TUOJIANGOSAURUS MULTISPINUS

JURASSIC

TITHONIAN	
KIMMERIDGIAN	LATE
OXFORDIAN	
CALLOVIAN	
BATHONIAN	MIDDLE
BAJOCIAN	
AALENIAN	
TOARCIAN	
PLIENSBACHIAN	EARLY
SINEMURIAN	
HETTANGIAN	

160mya

7m
(23ft)

H

HERBIVOROUS

4000kg
(3.9 tons)

SZECHUAN,
CHINA

This spectacular stegosaur is the best known to emerge from China, where it flourished in the Late Jurassic. Whereas the American *Stegosaurus* sported plates along its spine, this Asian genus possessed an array of sharp spikes that served to defend its back while it swung its even spikier tail at attackers. When it was not facing the unwelcome attentions of predators such as *Yangchuanosaurus*, it spent its time eating low foliage in dense woodlands. Known from two specimens found near the Tuo River in Szechuan province, it was named in 1977, a century after its more famous American relative.

105

FIELD NOTES FROM A PALAEONTOLOGIST

Jack Horner, world-leading palaeontologist, shares what it's like to find dinosaurs – and how you can get involved, too!

Dinosaur remains are common in North America, particularly in the West, along the eastern front of the Rocky Mountains from Alberta to Mexico. Collecting fossils, however, requires permission from the landowner. Every piece of land in the United States and Canada is owned by someone or some entity, so wherever you decide to go, it is important that you have permission to look for fossils. Whoever owns the land owns the fossils on the land.

If your interest lies more in *seeing* dinosaur fossils still embedded in the rock than in *finding* them yourself, there are two spots in the United States where you can view dinosaur specimens right where they were found by palaeontologists (a far more practical way to guarantee yourself a fossil sighting). One is Dinosaur National Monument, in the Uinta Mountains on the border between Colorado and Utah, which also features hundreds of dinosaur fossils in the Dinosaur Quarry exhibit hall. The other spot is Cleveland-Lloyd Quarry, in east central Utah, where more than 10,000 dinosaur bones represent the densest concentration of Jurassic-aged dinosaur bones on record.

But if you're set on searching for specimens on your own, here's the conundrum: You're most likely to find a dinosaur on land where it's illegal to dig. Most yet-to-be-discovered dinosaurs turn up in places where people cannot raise animals or crops, but this unusable land is generally owned by the federal government. In Canada, all vertebrate fossils belong to the province in which they're found, and private collecting is illegal. In the United States, private collecting of vertebrate remains is illegal on all federal- and state-owned lands, and therefore collecting can only be done on private lands with the landowner's permission. Most places where you can personally collect dinosaurs are ranches, typically with an arrangement where the landowner receives a fee.

By far a better bet – if you want to hunt for dinosaurs the way palaeontologists do – would be to attend a palaeontological field school. Various for-profit or non-profit organizations offer field schools where participants can pay a fee to learn palaeontological techniques and aid the organization in collecting specimens for education or profit. Museum-based field schools, led by scientists, generally offer the best experiences.

Finding a dinosaur skeleton requires either incredible luck or a combination of knowledge and patience. As a palaeontologist working at a museum, I occasionally get calls from people who accidentally find extraordinary specimens. These people may know little or nothing of palaeontology, but they recognize that a specimen they've stumbled upon might be something important, so they report what they've found. This kind of discovery, of course, is the lucky kind. But for the most part, dinosaur specimens are found by people who have studied geology and palaeontology and spend their time actively searching.

I am this sort of palaeontologist, and every year I send out exploration teams of students and researchers who hunt for dinosaur specimens. For the past 15 years, my teams have been searching on state and federal lands in eastern Montana for dinosaurs that died within three million years of the extinction event some 65 million years ago. These teams have discovered troves of dinosaur bones, including more than one hundred specimens of *Triceratops* and a dozen of *Tyrannosaurus rex*, which are all brought back to the Museum of the Rockies, where fossil preparators clean the sediment from the bones and prepare them for study and display.

But before the public gets a glimpse, my colleagues and students and I carefully examine our new finds and publish papers on our findings and theories. Then we design museum exhibits based on what we've learned. Today, the general public can visit the Museum of the Rockies to see the many specimens we've uncovered and learn about our discoveries.

Here's some incentive for readers who are interested in getting involved themselves: Amateurs *do* get credit when they make a discovery and then report it to us – in some cases they even help us out with the digging. One such individual who has taken part in unearthing his own finds is Ken Olson, a retired Lutheran minister. Ken has followed up on many of his dinosaur sightings and even done the initial excavations. Several specimens he uncovered are on display in the Museum of the Rockies.

I, too, have made a number of discoveries over the years. My favourite was back in 1984, when I found one of the first clutches of dinosaur eggs to contain embryonic skeletons. I'd been out wandering, looking around for fossils, when I came across what I recognised was a clutch of dinosaur eggs. I began to patiently sweep the dust away from the eggshells with a small paintbrush, and as I examined the eggshell fragments I noticed tiny little bones protruding from within the broken eggshells. Some of the bones were so small that they were hard to see without a magnifying glass. In all, the clutch consisted of 19 eggs, each containing the tiny skeleton of a meat-eating dinosaur called *Troodon*. It was one of the most exciting days of my life: I was looking at one of the very first dinosaur embryos that had ever been found anywhere in the world.

My discovery is an excellent example of one based on both luck and knowledge. I knew I was in the right place to find dinosaur eggs because I was walking through badlands terrain – a dry area where the rock and soil have been extremely eroded – within a rock unit called the Two Medicine Formation. Many years before I was there, geologists had determined that the Two Medicine Formation is made of sediment deposited by rivers between 80 million and 74 million years ago. Dinosaurs lived from 200 million years ago until 65 million years ago, so I knew the rock formation was the right age. I'd also previously searched for dinosaur bones nearby, and turned up baby bones and small pieces of eggshell, so I knew I might find dinosaur eggs in the area. I had the requisite knowledge to put myself in the right place among rocks from the right geological time. But because there is no rule as to exactly where a dinosaur might have died, even if you are in the right general area, finding the remains of dinosaurs requires a good deal of luck as well.

Finding a dinosaur – or any other fossil – is simply a matter of inspecting every piece of rock you can possibly find. This is accomplished by walking around and scrutinising the ground, hunting for little bits of bone or other fossil material. Some specimens are small fragments of large bones and some are whole bones; whole bones are always better than fragments because they give us much more information. And, unfortunately, when you find a bone sticking out of the rock,

there's no easy way to determine whether there is any more of the skeleton in the ground – you've got to dig. It can be a long process, and it takes the right tools and glues so that the specimen does not fall apart while you are exposing it.

The discovery in 2000 of a significant *Tyrannosaurus rex* specimen, dubbed 'B-rex' by my field crew, is emblematic of the chance, intuition, and teamwork that make for a successful excavation. B-rex was found by my field crew chief, Bob Harmon – the B is for Bob – who had been out searching for dinosaurs and hadn't found much by lunchtime, so he climbed up on a ledge and ate his sandwich. When he finished his meal, he looked at the very steep sandstone cliff behind him and recognised a bone sticking out of the cliff side. He couldn't quite reach it, so he walked to his camp and got a folding chair, which he took back to the site and placed on top of a pile of rocks. He climbed up and saw that the bone was a metatarsal, or foot bone, from a *T. rex*.

While he excavated the metatarsal, he realized there were a few more bones nearby in the same rock layer. Bob brought the foot bone to show me and explained that there were more bones visible in the cliff. I went out to the site and decided that we should excavate to see if there were still more bones, buried deeper. I brought in a crew of excavators capable of hanging on ropes over the side of the cliff while digging down to the potential skeleton. It was a delicate, time-consuming process that took several weeks, but when they got to the bone layer, they found the hind legs and parts of the skull.

It was clear to me then that this was going to be an important *Tyrannosaurus rex* specimen, so I instructed them to excavate a much larger area, and over the course of three summers the B-rex was uncovered. The excavators had to remove one thousand cubic yards of rock, all by hand. And because the specimen was far from any roads, all of the equipment used to excavate the specimen – and the bones we dug out – were transported by helicopter.

It was a very long, difficult excavation, but in the end B-rex turned out to be one of the most scientifically important dinosaur specimens ever collected. The bones were eventually determined to be from a young female, and they contained the first samples of dinosaur proteins and soft tissues ever found. One never knows which specimens will turn out to reveal new scientific information, so it is vital that they are all excavated very carefully, with the guidance of professionals, to preserve the precious clues that they hold.

Jack Horner with the B-rex femur

JURASSIC

TITHONIAN	
KIMMERIDGIAN	LATE
OXFORDIAN	
CALLOVIAN	
BATHONIAN	
BAJOCIAN	MIDDLE
AALENIAN	
TOARCIAN	
PLIENSBACHIAN	
SINEMURIAN	EARLY
HETTANGIAN	

155–150 mya

C

CARNIVOROUS

5kg (11lb)

GUIMAROTA, PORTUGAL

(AY-vee-ah-tie-RAN-us)

AVIATYRANNIS JURASSICA

MEANING 'TYRANT'S GRANDMOTHER'

Tyrannosaurs of the Late Jurassic were small and lithe, a far cry from their famous descendants around 80m years later. *Aviatyrannis*, a dog-sized European predator whose name means 'tyrant's grandmother', was one of the first tyrannosauroids along with the earlier *Guanlong* and *Proceratosaurus*, and its American contemporary *Stokesosaurus* – which is so similar that they may even be the same creature. *Aviatyrannis'* remains were found in Portugal and described in 2003, but they are so partial – only a couple of fragments of hip – that it is very hard to place accurately. The bones were found at a site called Guimarota, a disused coalmine more famous for its Jurassic mammal fossils than for dinosaurs' remains. This would have been part of a warm, dry, wooded island in the Late Jurassic. As well as the mammals that it probably hunted, *Aviatyrannis* shared this habitat with sauropods, troodontids and small theropods such as *Compsognathus* (page 127).

1m (3ft)

(CAMP-toe-SORE-us)

CAMPTOSAURUS DISPAR

· ·▶

5m
(16ft)

JURASSIC

TITHONIAN
KIMMERIDGIAN — LATE
OXFORDIAN
CALLOVIAN
BATHONIAN — MIDDLE
BAJOCIAN
AALENIAN
TOARCIAN
PLIENSBACHIAN — EARLY
SINEMURIAN
HETTANGIAN

150mya

H

HERBIVOROUS

500kg
(1100lb)

WYOMING,
USA

This bulky ornithopod ground its way through a diet of tough vegetation judging by the wear marks found on its fossilised teeth. Like the later iguanodontids to which it was related, it had shorter forelimbs with a spiked thumb, and could plod along on all fours or run on two legs at up to 15mph if necessary, for instance to escape a hungry *Allosaurus*. Othniel Charles Marsh named *Camptosaurus* in 1885, six years after its discovery in Wyoming, and further finds led to another 11 species joining the genus in the subsequent century. However, this is the only one still recognised, as the others were either younger specimens of this species or have become separate genera such as *Cumnoria* (page 186) and *Owenodon*.

111

(ORN-ith-oh-LEST-eez)

ORNITHOLESTES HERMANNI

One partial skeleton discovered in 1900 is all that is known of an agile little predator that scurried and pounced its way around the Jurassic woodlands. Its especially strong skull and conical teeth made it a more fearsome hunter than most dinosaurs of its size. Henry Fairfield Osborn granted it a name meaning 'bird robber' and while it may have caught Early Jurassic birds similar to *Archaeopteryx*, it probably subsisted on mammals, lizards, amphibians, baby dinosaurs and pterosaurs.

2m
(6ft 6in)

JURASSIC

TITHONIAN	LATE
KIMMERIDGIAN	
OXFORDIAN	
CALLOVIAN	MIDDLE
BATHONIAN	
BAJOCIAN	
AALENIAN	
TOARCIAN	EARLY
PLIENSBACHIAN	
SINEMURIAN	
HETTANGIAN	

154mya

CARNIVOROUS

11kg
(24lb)

WYOMING,
USA

113

TITHONIAN	
KIMMERIDGIAN	LATE
OXFORDIAN	
CALLOVIAN	
BATHONIAN	MIDDLE
BAJOCIAN	
AALENIAN	
TOARCIAN	
PLIENSBACHIAN	EARLY
SINEMURIAN	
HETTANGIAN	

154–151
mya

HERBIVOROUS

10,000kg
(9.8 tons)

TANZANIA

(yan-ENSH-ee-ah)

JANENSCHIA ROBUSTA

Think of the mighty titanosaurs such as *Argentinosaurus* and *Paralititan*, some of the heaviest land animals ever to have existed. Weighing a relatively meagre 9.8 tons and as long as a train carriage, *Janenschia* was the forerunner of them all, hailing from the Jurassic age whereas the other titanosaurs lived in the Cretaceous. They would go on to become prevalent across the world: their fossils have been found on every continent apart from Antarctica, and that is probably only because they lie inaccessible beneath the ice and snow.

up to 20m
(65ft)

JURASSIC

TITHONIAN	
KIMMERIDGIAN	LATE
OXFORDIAN	
CALLOVIAN	
BATHONIAN	MIDDLE
BAJOCIAN	
AALENIAN	
TOARCIAN	
PLIENSBACHIAN	EARLY
SINEMURIAN	
HETTANGIAN	

(yoo-ROPE-ah-SORE-us)

EUROPASAURUS HOLGERI

6m
(20ft)

155–150
mya

HERBIVOROUS

500kg
(1100lb)

The macronarians included some of the truly gigantic dinosaurs such as *Sauroposeidon* (page 206) – but this one is notable for being tiny, at least by sauropods' standards. It's likely that *Europasaurus*, found in northern Germany, is an example of 'island dwarfism', the phenomenon whereby animals evolve to shrink in size in order to cope with the diminished resources available. The area of Lower Saxony where its skull and a few vertebrae were found lay among a scattering of little islands in the Late Jurassic. Analysis of its bones shows that whereas the gigantic sauropods gained their great size by growing extremely quickly, *Europasaurus* had an unusually slow rate of growth. Several individuals have been found in the Langenberg Quarry spanning from 1.7m to 6.2m (5ft 7in to 20ft) in length, from which it is deduced that typical adults held their heads aloft 3m (10ft) in the air and, despite being relatives of *Brachiosaurus*, at their shoulders were only as high as a tall human.

LOWER
SAXONY,
GERMANY

115

JURASSIC

TITHONIAN	
KIMMERIDGIAN	LATE
OXFORDIAN	
CALLOVIAN	
BATHONIAN	MIDDLE
BAJOCIAN	
AALENIAN	
TOARCIAN	
PLIENSBACHIAN	EARLY
SINEMURIAN	
HETTANGIAN	

155mya

HERBIVOROUS

20kg
(44lb)

**WYOMING,
USA**

116

(DRINK-er)

DRINKER NISTI ·········· AND ··

In 1877 Othniel Marsh named a small, speedy hypsilophodont known from two skeletons found in the western USA. His preferred name was *Nanosaurus*, meaning 'little lizard', but a century later a British palaeontologist named Peter Galton honoured Marsh by renaming it *Othnielia*, which was more recently adapted to *Othnielosaurus*.

**2m
(6ft 6in)**

Galton was also involved in 1990 in the description of another lightweight running dinosaur. He and American expert Robert Bakker decided to even matters up between the old Bone Wars rivals by naming it *Drinker* for Edward Drinker Cope.

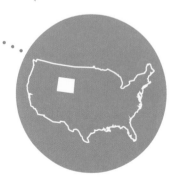

(oth-NEE-lee-oh-SORE-us)

OTHNIELOSAURUS CONSORS

2.2m
(7ft)

JURASSIC

TITHONIAN	
KIMMERIDGIAN	LATE
OXFORDIAN	
CALLOVIAN	
BATHONIAN	MIDDLE
BAJOCIAN	
AALENIAN	
TOARCIAN	
PLIENSBACHIAN	EARLY
SINEMURIAN	
HETTANGIAN	

155–145
mya

HERBIVOROUS

30kg
(66lb)

WYOMING,
COLORADO AND
UTAH, USA

117

Both little ornithopods would have flitted and scurried around the mighty sauropods they lived alongside, pecking at low-growing vegetation that was beyond their bigger contemporaries' reach. *Drinker* seems to have had a more flexible tail and spread-out toes that suggest it lived on soft ground, perhaps a swampy environment, but both dinosaurs are much alike.

In fact the similarities are so great that it has been suggested that they are the same creature. As the name established first always takes precedence, in this case *Drinker* would become *Othnielosaurus* – a little ironic given that Marsh also bested Cope in the Bone Wars (page 88).

TITHONIAN	
KIMMERIDGIAN	LATE
OXFORDIAN	
CALLOVIAN	
BATHONIAN	MIDDLE
BAJOCIAN	
AALENIAN	
TOARCIAN	
PLIENSBACHIAN	EARLY
SINEMURIAN	
HETTANGIAN	

153–148
mya

C

CARNIVOROUS

AS LONG
AS A KILLER
WHALE

1950kg
(1.9 tons)

COLORADO, USA
– AND MAYBE
PORTUGAL

118

(TORE-voh-SORE-us)

TORVOSAURUS TANNERI

This is one of many spectacular Jurassic dinosaurs extracted from the rich repository of bones within the Morrison Formation, a seam of sandstone and mudstone running through the American Midwest. The remains were washed to their resting place by a river that ran through a valley here around 150mya. *Torvosaurus'* only confirmed fossil, drawn from Colorado's Dry Mesa Quarry by James Jensen, is a single forelimb, but other remains are attributed including a piece of skull, a jawbone, a hip fragment and neck vertebrae. Combined, they suggest that this was the biggest carnivore of its time – as big as *Allosaurus* and heavier too, around the weight of a hippo. Its small but powerful arms had large claws, and a jawbone found in Portugal in 2006, suspected to be from a species of *Torvosaurus*, suggests its skull was up to 1.5m (5ft) long. Although its name means 'savage lizard', as with many of the big meat-eating dinosaurs a question lingers over whether it was actually a hunter or a scavenger.

9m
(30ft)

(MARSH-o-SORE-us)

MARSHOSAURUS BICENTESIMUS

JURASSIC

TITHONIAN	
KIMMERIDGIAN	LATE
OXFORDIAN	
CALLOVIAN	
BATHONIAN	MIDDLE
BAJOCIAN	
AALENIAN	
TOARCIAN	
PLIENSBACHIAN	EARLY
SINEMURIAN	
HETTANGIAN	

5m
(16ft)

154–142 mya

C

CARNIVOROUS

200kg
(440lb)

This fearsome predator had curved, serrated teeth set into a powerful, 60cm-long (2ft) skull. It was discovered in the Cleveland-Lloyd Quarry in Utah, a rich source of finds for Othniel Charles Marsh during the late 19th century. This wasn't one of them, however. *Marshosaurus* was named in his honour in 1976; as alluded to by its species name, *M. bicentesimus*, that year marked the bicentenary of American independence. Many *Allosaurus* fossils have been found in the same location, suggesting that the two vicious carnivores lived alongside each other.

POWERFUL 60CM-LONG SKULL

COLORADO AND UTAH, USA

JURASSIC

TITHONIAN	LATE
KIMMERIDGIAN	
OXFORDIAN	
CALLOVIAN	MIDDLE
BATHONIAN	
BAJOCIAN	
AALENIAN	
TOARCIAN	EARLY
PLIENSBACHIAN	
SINEMURIAN	
HETTANGIAN	

168–160 mya

?

INSECTIVOROUS?

0.2kg
(0.5lb)

INNER
MONGOLIA

120

(EPP-ee-dex-IP-ter-ix)
EPIDEXIPTERYX HUI

This strange little creature's discovery in 2008 changed the understanding of feathers' evolution. Here was a pigeon-sized flightless dinosaur covered in fluffy down and boasting four long, straight and very striking tail feathers that must have been ornamental – and it dated from between 168 and 160mya, in the Middle to Late Jurassic. That is significantly before *Archaeopteryx*, traditionally considered the first known bird, which dates from 147mya. So *Epidexipteryx* proves that feathers were being used for display for millions of years before they developed a flight purpose. The appearance is reminiscent of some modern birds of paradise, and they may have had a similar dual use of attracting mates and threatening enemies.

The exquisite fossil came from the Daohugou beds in Nincheng County, Inner Mongolia, which is a semi-autonomous province of China. As well as the distinct feather impressions it showed small claws and a long third finger, which could have been used for extracting beetle grubs from bark as it scampered up and down tree trunks.

30cm
(1ft)

TITHONIAN	
KIMMERIDGIAN	LATE
OXFORDIAN	
CALLOVIAN	
BATHONIAN	MIDDLE
BAJOCIAN	
AALENIAN	
TOARCIAN	
PLIENSBACHIAN	EARLY
SINEMURIAN	
HETTANGIAN	

159–142 mya

H

HERBIVOROUS

226kg (500lb)

SZECHUAN PROVINCE, CHINA

(jah-LING-oh-sore-us)

CHIALINGOSAURUS KUANI

This was the first stegosaur discovered in China and one of the earliest to have lived. *Chialingosaurus* was smaller and slenderer than most of its descendants: the partial skeleton found at the Chialing River in 1959 shows that it had a high and narrow skull, a few well-spaced teeth and a series of relatively small plates and spikes running in two rows along its back. However, it's possible that its small size is due to *Chialingosaurus* actually being a juvenile *Tuojiangosaurus*, which grew far larger and bulkier. It would have eaten low-growing fern-like plants, which were abundant in the Jurassic, but its tail vertebrae suggest it might have been able to rear up on its hind legs, in order to browse on tree foliage.

4m (13ft)

(SOO-per-SORE-us)
SUPERSAURUS VIVIANAE ········▶

Supersaurus had the longest neck of any dinosaur known – it stretched 15m (49ft) from its shoulders to its skull, about the length of one-and-a-half double-decker buses. We know this as it is one of the biggest dinosaurs to have left us a near-complete skeleton, a cast of which stands on display at the Wyoming Dinosaur Center in the USA. Standing approximately 34m (112ft) long, this was an even larger relative of the better known *Diplodocus*. It was discovered amid the clutter of Late Jurassic sauropod fossils at Colorado's Dry Mesa Quarry. The site holds a notorious jumble of old bones: it is thought that a drought led thousands of dinosaurs to die of starvation, and then a flash flood washed their carcasses to this spot, where they sank into mud and became fossilised.

LONGEST NECK OF ANY KNOWN DINOSAUR

34m (112ft)

JURASSIC

TITHONIAN	
KIMMERIDGIAN	LATE
OXFORDIAN	
CALLOVIAN	
BATHONIAN	MIDDLE
BAJOCIAN	
AALENIAN	
TOARCIAN	
PLIENSBACHIAN	EARLY
SINEMURIAN	
HETTANGIAN	

153mya

HERBIVOROUS, HIGH OR LOW BROWSER

35,000kg (34.4 tons)

USA

JURASSIC

TITHONIAN	LATE
KIMMERIDGIAN	
OXFORDIAN	
CALLOVIAN	MIDDLE
BATHONIAN	
BAJOCIAN	
AALENIAN	
TOARCIAN	EARLY
PLIENSBACHIAN	
SINEMURIAN	
HETTANGIAN	

150mya

H

HERBIVOROUS

**2300kg
(2.2 tons)**

**USA,
PORTUGAL
AND PERHAPS
CHINA**

124

(steg-oh-SORE-us)

STEGOSAURUS ARMATUS

'The Museum of Yale College has recently received the greater portion of the skeleton of a huge reptile, which proves to be one of the most remarkable animals yet discovered…'

So wrote Othniel Marsh in the *American Journal of Science* in 1877, after one of his employees working in Colorado's Morrison Formation excavated some vertebrae, limbs, hipbones and a few bizarre spade-like plates, one measuring over a metre in length. Marsh was right – but it took a while for the great dinosaur-finder to figure out just what form these remarkable remains ought to take.

In naming it *Stegosaurus* ('roofed lizard'), initially he described the plates lying flat like roof-tiles and imagined that they formed the shell of a giant turtle. Then in 1879 after obtaining further remnants including a piece of skull, he concluded it was a dinosaur, but clung to the idea that it was aquatic. Then in 1885 he had the breakthrough he needed when a near-perfect specimen was discovered, again in Colorado. It had been fossilised lying on its side and crushed flat, leading it to be dubbed the 'roadkill *Stegosaurus*'. The plates that Marsh had pictured fitting together into a carapace were preserved in two overlapping rows protruding upright from its spine. When Marsh attempted a reconstruction he conjured an

image familiar to anyone who has heard of *Stegosaurus* – which is to say anyone with the vaguest interest in dinosaurs.

He portrayed a low-slung head and a stocky neck that rises into a great curved back, which descends again into a drooping tail tipped with sharp spikes pointing upward. This is the picture most of us grew up knowing, but now experts portray *Stegosaurus* with the spikes facing sideways and backward, and the tail held in a stiff horizontal posture about 1.8m (6ft) off the ground – high enough for most humans to stand beneath. Incidentally stegosaur spikes have a great name: they're known as a thagomizer, a word that originated in one of Gary Larson's The Far Side cartoons and then passed into general usage.

Lingering confusion has surrounded the purpose of its plates. They were too delicate to be of much use as protection and were only attached to the body by cartilage rather than bone, which is why *Stegosaurus* skeletons often have their plates missing or displaced. Nor could they have served to regulate temperature. Blood vessels ran around the plates, so it has been suggested that cool air blowing around them would cool the blood and hence the body temperature – but its close relative *Kentrosaurus* had spikes rather than plates

9m
(30ft)

on its back, and we would expect two related animals to possess the same cooling system. The current favoured theory is that they were for identification among fellow *Stegosaurus* and maybe for sexual display: blood pumping around them could cause them to 'blush', helping to attract a mate.

Stegosaurus gave its name to the stegosaurs, the family of related dinosaurs that lived throughout the Jurassic period and into the Early Cretaceous; their fossils have been found in Europe, North America, Africa and Asia. Indeed *Stegosaurus* itself may have endured into the Cretaceous in China according to a study published in 2008 by Susannah Maidment and colleagues, who suggest that the stegosaur *Wuerhosaurus* was actually a specimen of this dinosaur. *Stegosaurus* was among the biggest members of this spectacular family and continues to justify its fame as 'one of the most remarkable animals yet discovered'.

(comp-sog-NAYTH-us)

COMPSOGNATHUS LONGIPES ········▶

JURASSIC

TITHONIAN		
KIMMERIDGIAN		LATE
OXFORDIAN		
CALLOVIAN		
BATHONIAN		MIDDLE
BAJOCIAN		
AALENIAN		
TOARCIAN		
PLIENSBACHIAN		EARLY
SINEMURIAN		
HETTANGIAN		

The early discoveries of the 19th century kept revealing huge creatures – and then in 1861 *Compsognathus* emerged to show how vastly dinosaurs varied in size. For a long time this was famous as the smallest known non-avian theropod. While this position has since been usurped by tiny carnivores such as *Microraptor*, it retains an important place in the dinosaur family tree. Three years after its discovery in Germany an anatomist named Carl Gegenbaur observed the similarity between *Compsognathus*' ankle bones and modern birds'. In doing so he became the first person to raise the possibility of dinosaurs' development into birds – so *Compsognathus* stands at the very beginning of scientists' efforts to understand this evolutionary link.

Relatives such as *Sinosauropteryx* had a covering of protofeathers but the *Compsognathus* fossils reveal no hint of this, so its external appearance is uncertain. It was a long-legged nimble runner with grasping claws and tiny sharp teeth set in a delicate skull. Its name means 'dainty jaw'.

Another reason for many people's continued familiarity with *Compsognathus* is its inclusion in the *Jurassic Park* films, where it was depicted hunting in packs. There is no evidence to suggest that it was anything but a lone hunter, and its prey may have included *Archaeopteryx*; the two creatures' fossils were recovered from the same German limestone formation. A second well-preserved and significantly larger *Compsognathus* fossil turned up in France in 1972, leading to suggestions that the German specimen was actually a juvenile.

150mya

CARNIVOROUS

2.5kg
(5.5lb)

SOUTHERN GERMANY AND FRANCE

1.25m
(4ft)
· · · · · · · · · · · · · ▶

127

JURASSIC

TITHONIAN	LATE
KIMMERIDGIAN	
OXFORDIAN	
CALLOVIAN	MIDDLE
BATHONIAN	
BAJOCIAN	
AALENIAN	
TOARCIAN	EARLY
PLIENSBACHIAN	
SINEMURIAN	
HETTANGIAN	

154–150
mya

HERBIVOROUS

27,000kg
(26.5 tons)

WYOMING,
UTAH AND
COLORADO,
USA

128

(a-PAT-oh-SORE-us)
APATOSAURUS EXCELSUS

In 1877 Othniel Charles Marsh found a large sauropod at Como Bluff in Wyoming and named it *Apatosaurus*. Two years later he found the largest dinosaur yet known and gave it a fitting name: *Brontosaurus excelsus*, meaning 'thunder lizard'.

However, in 1903 further study showed that the two animals were different species of the same genus. The rules of scientific classification dictated that the older name took precedence – but '*Brontosaurus*' is so evocative of this massive creature's imposing stature that the name remains familiar, more than a century after it was discarded from scientific use. Judging by the fossils obtained from the rich Morrison Formation, *Apatosaurus* was the second commonest sauropod in the western USA during the Late Jurassic, after *Camarasaurus* (page 100). It belonged to the diplodocid family but was far bulkier than its relative *Diplodocus* and typically weighed in at around 30 tons, around the same as seven elephants. *Apatosaurus* must have been an insatiable herbivore but questions remain about how it fed; some experts believe that diplodocids'

necks were too inflexible to raise to the treetops and suggest that *Apatosaurus* used its incredible strength to knock trees down to the ground, where it could browse from them more easily. Others disagree strongly with this idea of limited neck mobility and argue that *Apatosaurus* and its relatives were almost certainly capable of raising the neck high up, and of sweeping it far to the side when foraging.

Footprints of a young *Apatosaurus* found in Colorado in 2006 supposedly show that these dog-sized juveniles could run on their hind legs, just as basilisk lizards do today. This would have helped them flee predatory theropods, for which baby sauropods represented an easier meal than their dangerously powerful parents. Adult *Apatosaurus* probably

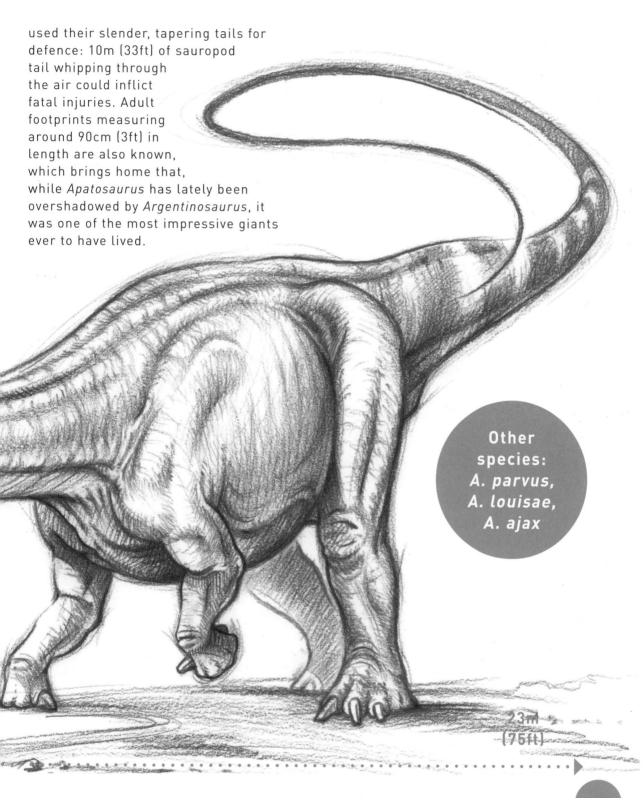

used their slender, tapering tails for
defence: 10m (33ft) of sauropod
tail whipping through
the air could inflict
fatal injuries. Adult
footprints measuring
around 90cm (3ft) in
length are also known,
which brings home that,
while *Apatosaurus* has lately been
overshadowed by *Argentinosaurus*, it
was one of the most impressive giants
ever to have lived.

**Other
species:
*A. parvus,
A. louisae,
A. ajax***

23m
(75ft)

JURASSIC

TITHONIAN	
KIMMERIDGIAN	LATE
OXFORDIAN	
CALLOVIAN	
BATHONIAN	MIDDLE
BAJOCIAN	
AALENIAN	
TOARCIAN	
PLIENSBACHIAN	EARLY
SINEMURIAN	
HETTANGIAN	

155–145
mya

HERBIVOROUS

23,000kg
(22.6 tons)

TENDAGURU,
TANZANIA

130

(ji-RAFF-oh-tie-tan)

GIRAFFATITAN BRANCAI .... AND ...

One of the most famous sauropods' confusing tale began in 1900 when fossil-hunter Elmer Riggs found *Brachiosaurus altithorax*'s giant bones in the USA. Then in 1914 the German palaeontologist Werner Janensch found similar but better remains in Tanzania that he deemed another species and called *Brachiosaurus brancai*. This latter species formed our picture of *Brachiosaurus*, which was for many years the biggest dinosaur confirmed. Familiar depictions show *Brachiosaurus* with a giraffe-like physique, a long neck forming half its body-length, a domed cranium and protruding jaws, a ridged backbone and an unusually short tail for a sauropod... but in 1988 American palaeontologist Gregory Paul concluded that the species *B. brancai* was actually a different genus altogether. When we think of *Brachiosaurus*, what

we picture is the animal he named *Giraffatitan*. Therefore since the split into two genera we have a far weaker understanding of *Brachiosaurus*' appearance.

It took another 21 years for his controversial reclassification to be accepted, though. Paul was vindicated in his attempt to recategorise this well-established dinosaur in 2009 when British palaeontologist Mike Taylor scrutinised and compared every type of bone common to both animals' fossils, and confirmed that 26 had distinct differences.

Extrapolating from the few bones that are still deemed *Brachiosaurus* – some back vertebrae, a couple of tail vertebrae, some legbones and ribs – Dr Taylor judges it to be a significantly bigger animal: the size and shape of the vertebrae suggest its torso and tail were

22.5m
(73ft)

(BRAK-ee-oh-SORE-us)

BRACHIOSAURUS ALTITHORAX

25m
(82ft)

JURASSIC

TITHONIAN	LATE
KIMMERIDGIAN	
OXFORDIAN	
CALLOVIAN	
BATHONIAN	MIDDLE
BAJOCIAN	
AALENIAN	
TOARCIAN	
PLIENSBACHIAN	EARLY
SINEMURIAN	
HETTANGIAN	

153mya

HERBIVOROUS

28,000kg
(27.5 tons)

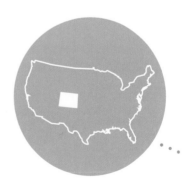

COLORADO, USA

131

both around a quarter longer than *Giraffatitan*'s. It also seems to have had a bulkier body and even longer front limbs that may have sprawled outwards. What's more, the lack of fusing between some of *Brachiosaurus*' bones suggests it wasn't even fully grown.

Both animals were huge, hungry herbivores, eating up to an estimated 120kg (260lb) of cycads, conifer and gingko leaves every day. Their long front legs allowed them to reach higher leaves than typical sauropods, which had limbs of roughly equal length and a horizontal posture. However, unlike other sauropods they couldn't gain extra height by rearing up on their hind legs. Their centre of gravity was too far forward and the front limbs too weak to withstand the force of their great bulk crashing back down to the ground.

JURASSIC

TITHONIAN	
KIMMERIDGIAN	LATE
OXFORDIAN	
CALLOVIAN	
BATHONIAN	MIDDLE
BAJOCIAN	
AALENIAN	
TOARCIAN	
PLIENSBACHIAN	EARLY
SINEMURIAN	
HETTANGIAN	

153–148 mya

CARNIVOROUS, EATING MEAT AND FISH

700kg (1500lb)

COLORADO, AND UTAH, USA

132

(SERR-at-oh-SORE-us)
CERATOSAURUS NASICORNIS

MOST DRAGON-LIKE DINOSAUR KNOWN

With a horn on its nose, two smaller lumps above its eyes and a covering of bony bumps along its back (an armour-plating unique among theropods), *Ceratosaurus* is the most dragon-like dinosaur known. This relative of the abelisaurs lived alongside *Allosaurus* and *Torvosaurus*: they probably focused on hunting the biggest sauropods, while *Ceratosaurus* may have combined aquatic hunting with picking off the smaller land-based herbivores. What's now the arid southwestern USA was then a swampy region and its deep and particularly flexible tail may have helped *Ceratosaurus* to scull through water in pursuit of crocodilians and fish. However, the tooth-marks it left on sauropod bones confirm that it would eat anything it could, though these larger animals may have been scavenged.

Its armour may have helped defend against bigger predators, but its horns were probably just used for display.

7m (23ft)

(JURE-a-ven-AYT-or)
JURAVENATOR STARKI

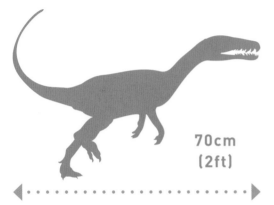

**70cm
(2ft)**

NOCTURNAL HUNTER

CARNIVOROUS

**1.5kg
(3.3lb)**

GERMANY

JURASSIC

TITHONIAN	
KIMMERIDGIAN	LATE
OXFORDIAN	
CALLOVIAN	
BATHONIAN	MIDDLE
BAJOCIAN	
AALENIAN	
TOARCIAN	
PLIENSBACHIAN	EARLY
SINEMURIAN	
HETTANGIAN	

152mya

After this little carnivore's discovery in southern Germany's Jura Mountains (which gave their name to the Jurassic period) its fossil was nicknamed Borsti, a common German name for bristle-haired dogs, because it was assumed that like its relative *Sinosauropteryx* it would have been covered by bristly proto-feathers. Further examination threw this into doubt: the fossil also preserved patches of skin impression around the tail and legs with small rounded scales. It was not until 2010, when scientists carried out a thorough analysis under ultraviolet light, that a fuller picture emerged. *Juravenator*'s fossil yielded the faintest traces of feather filaments, proving the early predictions correct, but also further scale impressions, which shows that 'dino-fuzz' existed in tufts or patches alongside a mostly scaly covering. The single specimen known was a juvenile – some of its bones had not fused as they would in an adult – so it is hard to know how much bigger it would have grown. Research in 2011 showed that *Juravenator*'s scleral rings (the circle of bones around the eye, controlling its contraction) would have enabled good vision in low light, suggesting it to be a nocturnal hunter. Its large teeth mean it could have preyed on relatively large animals, and a notch in its jaw suggests it was also adept at catching fish.

EGGS

All around our planet today lie thousands of eggs that will never hatch.

Some are longer than a rugby ball, others not much bigger than chickens' eggs, some arranged in circles and others in mounds. All have remained static for more than 65.5m years since the moment when powerful desert winds covered them in deep sand-drifts or floods immersed them in silt, but they can still tell us much about the creatures that laid them.

One of the most significant dinosaur nesting sites was found in the Rocky Mountains of Montana, USA, and described by American palaeontologist Jack Horner. In the Late Cretaceous a herd of duck-billed hadrosaurs settled here to lay their eggs – but also to stay and raise their young. Horner knew that the site, dubbed 'Egg Mountain', contained numerous shallow bowl-shaped nests that were 1.8m (6ft) in diameter and filled with up to 20 eggs each – but also with the fossils of hatchlings and baby dinosaurs up to two months old. Signs of regurgitated plant matter also suggested that the parents had brought food to the nests for their offspring. He named the hadrosaur *Maiasaura*, the 'good mother lizard', because it was clear that these dinosaurs cared for their babies rather than dropping their eggs and abandoning them. Other

dinosaurs did that, however, and turtles still do.

The *Maiasaura* site is important but the most spectacular is down in South America. In 1997 a team of American and Argentine palaeontologists was searching for dinosaur bones in a patch of the Patagonian badlands known as Auca Mahuevo. Then they realised that the very ground they were walking on was littered with what looked like cobble stones...

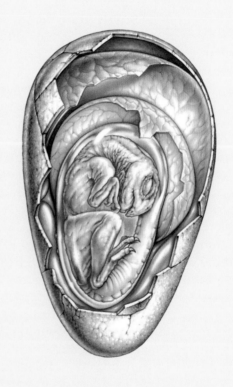

sauropod eggs, everywhere they looked across a square mile of arid landscape, many of them perfectly preserved after 80m years. These eggs remained in nests lying two to three metres (up to 10ft) apart that had been hollowed a few inches into the ground and insulated with plant material.

At another South American site, Sanagasta Geological Park in Argentina, scientists found fossil eggs buried in soil by the mother sauropods. The fossilised nests lay only a few feet away from certain crystals and mud layers denoting the presence of geysers and steaming vents. These natural sources of hot water would have warmed the damp soil, making it a perfect incubator for large eggs, which needed heat and moisture for the embryos to develop.

The first known dinosaur eggs were found in France in the 1870s but this

discovery received little publicity. Far more widely reported was the find made by Roy Chapman Andrews during his expedition to the Gobi Desert in the 1920s. His party found what they identified as a clutch of *Protoceratops* eggs – though in the 1990s it became apparent they belonged to *Oviraptor*. A *Protoceratops* discovery in Mongolia in 2011 was just as significant: a nest containing the finely preserved remains of 15 tiny ceratopsians proved that it wasn't only hadrosaurs and sauropods that looked after their offspring.

Striking evidence of the fact that predators saw baby dinosaurs as easy meat came to light in western India in 2011. A 67m-year-old nest of titanosaur hatchlings was discovered complete with the fossil of a 3.5m-long (11ft 6in) snake coiled around a crushed egg, and beside it the partial remains of a hatchling. Just as the snake – named *Sanajeh indicus* – was about to finish its meal, a landslide covered the whole area with mud.

From the biggest – those of the almighty *Gigantoraptor*, which laid tubular-shaped eggs up to 45cm (18in) long – to the smallest, dinosaur eggs are like other trace fossils such as footprints and coprolites, in that they can reveal so much about how the animals actually lived rather than only preserving the moment when they died.

JURASSIC

TITHONIAN	
KIMMERIDGIAN	LATE
OXFORDIAN	
CALLOVIAN	
BATHONIAN	MIDDLE
BAJOCIAN	
AALENIAN	
TOARCIAN	
PLIENSBACHIAN	EARLY
SINEMURIAN	
HETTANGIAN	

154–150 mya

C

CARNIVOROUS

3000kg (2.9 tons)

TANZANIA, AFRICA

136

(VET-er-oo-PRIST-ee-SORE-us)

VETERUPRISTISAURUS MILNERI

MEANING 'OLD SHARK LIZARD'

Sharks' jaws and carcharodontosaurs' jaws were very similar, except that the latter were far bigger; this African specimen is the earliest known, hence its name meaning 'old shark lizard'. A single tail vertebra found in Tanzania's famous Tendaguru rock formation was sufficient for German palaeontologist Oliver Rauhut to announce in 2011 that the carcharodontosaur family stretched back into the Late Jurassic (more famous members such as *Giganotosaurus* (page 239) lived in the Late Cretaceous). Two more tail vertebrae later found nearby have since been referred to this genus. The original vertebra was 12.3cm (3¾in), from which Rauhut deduced that this bipedal predator was between 8.5m and 10m (28–33ft) long – not a giant like its descendants but still a mighty predator that, in this specimen's case, stalked the muddy banks of a lagoon in Middle Jurassic eastern Africa. The fossil may have been a juvenile, in which case an adult *Veterupristisaurus* would have been bigger still.

10m (33ft)

(SIGH-oor-uh-MEE-mus)

SCIURUMIMUS ALBERSDOERFERI

Perhaps 6m (20ft) as an adult

JURASSIC

TITHONIAN	LATE
KIMMERIDGIAN	
OXFORDIAN	
CALLOVIAN	MIDDLE
BATHONIAN	
BAJOCIAN	
AALENIAN	
TOARCIAN	EARLY
PLIENSBACHIAN	
SINEMURIAN	
HETTANGIAN	

150mya

CARNIVOROUS C

uncertain

PAINTEN, GERMANY

Since the mid-1990s more than 30 genera of dinosaurs have been found with traces of feathers or a fuzzy covering. Most are coelurosaurs – the group that among others includes tyrannosaurs, deinonychosaurs and birds – though bristles and tufts are also known from primitive ornithischians such as *Psittacosaurus* (page 183). The 71cm-long (28in) juvenile *Sciurumimus'* discovery in Bavarian limestone in 2012 suggested that 'dino-fuzz' may have been far more widespread. Rather than a coelurosaur, this was a megalosauroid, a very different group of more primitive theropods. The fossil, probably the most perfectly preserved found in Europe, retains signs of this fuzz on its squirrel-like tail and in patches on the body. The bristles in this saurischian dinosaur seem structurally similar to those known from ornithischians, suggesting the possibility that both inherited the trait from a common ancestor before the dinosaurs' division into

their two main orders. When he and his colleagues named and described *Sciurumimus* (meaning 'squirrel mimic'), palaeontologist Oliver Rauhut argued that this means plumage was at least present at the base of the theropod family tree – so instead of being the preserve of the creatures that evolved into modern birds, the implication is that feathers of a sort were a very ancient and widespread feature in dinosaurs. More feathered fossils from outside the coelurosaurs will help consolidate the growing suspicion that throughout the dinosaurs, some kind of fuzz was the norm rather than the exception.

EARLY EVIDENCE OF 'DINO-FUZZ'

137

JURASSIC

TITHONIAN	
KIMMERIDGIAN	LATE
OXFORDIAN	
CALLOVIAN	
BATHONIAN	MIDDLE
BAJOCIAN	
AALENIAN	
TOARCIAN	
PLIENSBACHIAN	EARLY
SINEMURIAN	
HETTANGIAN	

155–145 mya

H

HERBIVOROUS

300kg (660lb)

COLORADO, USA

138

(my-MOOR-ah-PELT-ah)
MYMOORAPELTA MAYSI

3m (10ft)

This squat, armoured plant-eater is an early member of the ankylosaurs, the low-slung, almost indestructible creatures that thrived throughout the Late Cretaceous. The hallmarks are all here – a dense skull, impenetrable scutes all over its back, sharp spikes jutting out from its sides – but like its close relation *Gargoyleosaurus* (page 143), *Mymoorapelta* was relatively small. Later genera such as *Tarchia* would grow almost three times as long.

It takes its name from the Mygatt-Moore Quarry in Colorado where its scrambled remains lay within Morrison Formation rocks. *Mymoorapelta* was first described in 1994 but its placement within the ankylosaur family remained uncertain for many years. Then in 2011 a new analysis suggested it to be a basal member of the nodosaurids, the group of ankylosaurs that had spikes but didn't have tail clubs.

TITHONIAN		
KIMMERIDGIAN		LATE
OXFORDIAN		
CALLOVIAN		
BATHONIAN		MIDDLE
BAJOCIAN		
AALENIAN		
TOARCIAN		
PLIENSBACHIAN		EARLY
SINEMURIAN		
HETTANGIAN		

(DIP-lo-DOCK-us)
DIPLODOCUS LONGUS · · · · · · · · · · · · ▶

Anyone who has visited London's Natural History Museum will know *Diplodocus'* impressive proportions. The 26m-long (83ft) replica skeleton that greets visitors as they enter the Central Hall was given to the museum by King Edward VII in 1905. Three years earlier he had seen a detailed sketch of a *Diplodocus'* skeleton at Skibo castle in Scotland, the home of his friend Andrew Carnegie, the Scottish-born American philanthropist after whom the second known *Diplodocus* species was named. The King wanted his own *Diplodocus*, and Carnegie made enquiries on his behalf among palaeontologist friends. The reply came that finding the King another *Diplodocus* to order would be a little tricky, but they could create him a replica by taking casts from existing fossils. So that is what they did. 'Dippy', as the museum's replica has become known, combines precise remakes of 324 bones taken from three dinosaurs discovered in Wyoming, USA. It was unveiled before the elite of London society and soon became a great attraction for the general public. Whatever their station in Edwardian society, it was for most people the first complete reconstruction of a dinosaur skeleton they had seen – and the same remains true for many Britons today. It has occupied its present location in the museum since 1979, the only subsequent alteration coming in 1993 when its long tail, which comprises 80 vertebrae, was lifted from the ground to a horizontal position. *Diplodocus* lived in the semi-arid conditions of Late Jurassic North America. Its unusual arrangement of teeth suggests it could strip a branch of leaves with its mouth, rather than biting off clumps of foliage. Although it was one of the longest dinosaurs known, *Diplodocus* was slender compared with related sauropods such as *Apatosaurus*, though it would still have posed quite a challenge to predators such as *Allosaurus*. It may have used its whip-like tail to defend itself, and it had a single claw on its elephantine feet that probably served the same purpose. A fossilised skin impression found in Wyoming's Howe Quarry in 1990 also revealed that the tail carried a row of spines, which it is suspected ran all the way up the vertebrae to the neck. They were made of keratin, like a rhino's horn or your hair and fingernails, which explains why there is relatively little evidence in the fossil record.

typically 22–24m (72–79ft) but some up to 35m (115ft)

154–150 mya

HERBIVOROUS, BROWSER AND GRAZER

18,000kg (17.7 tons)

COLORADO, UTAH AND WYOMING, USA

BREAKING THE SOUND BARRIER

Long, whip-like tails were a distinctive feature of the diplodocoids – but what was their purpose? The Canadian palaeontologist Phil Currie and his colleague Nathan Myhrvold came up with a novel proposal. They calculated that a diplodocoid could swing its tail fast enough for the tip to exceed the sound barrier, creating a loud cracking noise just like a bullwhip. This could serve to warn off enemies and maybe also feature in a courtship display. Swinging its tail could also serve to to scare away, knock over or even kill predatory therapods.

JURASSIC

TITHONIAN	
KIMMERIDGIAN	LATE
OXFORDIAN	
CALLOVIAN	
BATHONIAN	MIDDLE
BAJOCIAN	
AALENIAN	
TOARCIAN	
PLIENSBACHIAN	EARLY
SINEMURIAN	
HETTANGIAN	

150–145 mya

HERBIVOROUS, GRAZER OR LOW BROWSER

5500kg (5.4 tons)

ARGENTINA

(BRACK-ee-tra-KEL-oh-pan)

BRACHYTRACHELOPAN MESAI

From the shoulders back to the tip of its long, thin tail, *Brachytrachelopan* had proportions typical of a sauropod: a fairly small one but still a familiar form. This makes its stubby little neck appear all the stranger, looking as if it were transplanted from another dinosaur altogether.

When *Brachytrachelopan*'s discovery was announced in 2005, the experts who described it argued that this weird neck probably helped it to eat some form of vegetation in Late Jurassic Argentina that grew to around a metre or two high and was particularly abundant or nutritious. The vertebrae suggest that it could not lift its neck higher than horizontal, preventing it from browsing the treetops. This specialised diet may also explain its relatively small size. Like *Amargasaurus* and *Dicraeosaurus* it would have had long pairs of spines along its neck, back and tail. For this reason the three are known as dicraeosaurids, meaning 'two-forked lizard'. *Brachytrachelopan*'s species name refers to a shepherd named Daniel Mesa, who found its eroding bones while searching for a lost sheep.

10m (33ft)

(gar-GOY-lee-o-sore-us)

GARGOYLEOSAURUS PARKPINORUM

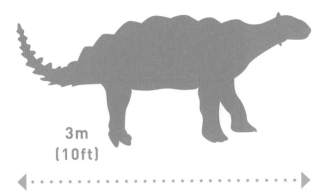

3m
(10ft)

This early ankylosaur's typically tough armour-coating formed a defence from predators such as *Allosaurus*. A carnivore's only chance of killing *Gargoyleosaurus* would have been to flip it on to its back and attack its vulnerable underbelly, as the combination of bony plates and sharp spikes protecting its body made it impenetrable from other angles. First described in 1998, this squat and slow-moving creature has a name derived from its skull's resemblance to the stone gargoyles that decorate gothic churches. Its holotype – the definitive example that enabled it to attain a formal classification – was discovered two years earlier at the Bone Cabin Quarry in Wyoming, USA. Later ankylosaurs would grow to up to 7m (23ft) long but *Gargoyleosaurus* was relatively diminutive: think of it as a prototype for the great tanks that were later to come.

JURASSIC

TITHONIAN	
KIMMERIDGIAN	LATE
OXFORDIAN	
CALLOVIAN	
BATHONIAN	MIDDLE
BAJOCIAN	
AALENIAN	
TOARCIAN	
PLIENSBACHIAN	EARLY
SINEMURIAN	
HETTANGIAN	

155–150
mya

HERBIVOROUS,
LOW GRAZER

1000kg
(0.9 tons)

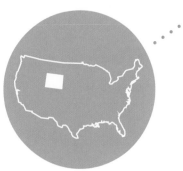

WYOMING,
USA

JURASSIC

TITHONIAN	
KIMMERIDGIAN	LATE
OXFORDIAN	
CALLOVIAN	
BATHONIAN	
BAJOCIAN	MIDDLE
AALENIAN	
TOARCIAN	
PLIENSBACHIAN	
SINEMURIAN	EARLY
HETTANGIAN	

150–145 mya

C

CARNIVOROUS

4500kg
(5 tons)

OKLAHOMA,
POSSIBLY
NEW MEXICO,
USA

(SORE-oh-FAG-ah-nax)

SAUROPHAGANAX MAXIMUS

The Morrison Formation's many layers provide a series of snapshots of Jurassic America, documenting its changing landscape and inhabitants over 10m years. Ancient floodplains, riverbeds and coverings of volcanic ash now form sharply delineated bands of rock in rugged badlands across the western to central USA.

If we assume that the countless dinosaurs lying within the formation – often recovered from sites such as Dry Mesa Quarry in Colorado and Como Bluff in Wyoming – were fossilised in rough proportion to their prevalence in life, *Saurophaganax* must have been a relatively rare sight in the Late Jurassic. Whereas the rocks have yielded dozens of the sauropod *Camarasaurus*, there are only vague hints of this enormous predator. It was an allosaurid comparable in size to *Tyrannosaurus*, sharing its close

relative *Allosaurus*' hugely muscled neck and slender, needle-sharp teeth. Its name means 'king of the lizard eaters' and in the Jurassic American woodlands it was certainly that, with *Apatosaurus* among its likely victims. A partial skeleton was collected in 1931 in Oklahoma (which has *Saurophaganax* as its official state fossil), and an even larger possible specimen from New Mexico awaits scientific analysis.

The Morrison spans a 600,000-square-mile swathe of the USA but only the very edges are visible: the rest lies buried beneath the Midwestern prairies, which must conceal an incredible assortment of unknown animals. Palaeontologists need another *Saurophaganax* to emerge before they can clearly picture one of the Jurassic's biggest and most brutal killers.

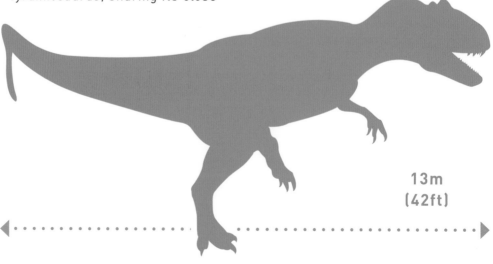

13m
(42ft)

(FROOT-ah-dens)
FRUITADENS HAAGARORUM

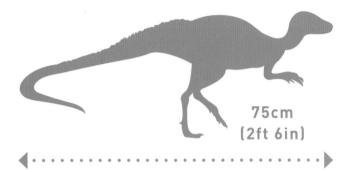

75cm
(2ft 6in)

It didn't eat fruit – there weren't any in the Jurassic or any flowers for that matter, as plants didn't evolve that reproduction strategy until the Cretaceous period. *Fruitadens* was the smallest known ornithischian and one of the very last of the heterodontosaurids (mixed-teeth dinosaurs), and is named for the city of Fruita in Colorado where its fossils were found in sandstone in the 1970s and 1980s. *Fruitadens* was a lean, running biped that used its mixture of teeth to eat anything: it had big sharp canines for biting into meat, and little grinding pegs for chewing vegetation. *Fruitadens'* jaws were lighter, quicker and opened wider than those of earlier heterodontosaurids – making them better adapted for snapping up small lizards and insects, while remaining able to chew plant matter too.

JURASSIC

TITHONIAN		
KIMMERIDGIAN	LATE	
OXFORDIAN		
CALLOVIAN		
BATHONIAN	MIDDLE	
BAJOCIAN		
AALENIAN		
TOARCIAN		
PLIENSBACHIAN	EARLY	
SINEMURIAN		
HETTANGIAN		

150mya

OMNIVOROUS

750g
(1.7lb)

COLORADO,
USA

BABY SAUROPODS

Baby sauropods running on two legs at the feet of their massive, four-legged parents.

JURASSIC

TITHONIAN	
KIMMERIDGIAN	LATE
OXFORDIAN	
CALLOVIAN	
BATHONIAN	MIDDLE
BAJOCIAN	
AALENIAN	
TOARCIAN	
PLIENSBACHIAN	EARLY
SINEMURIAN	
HETTANGIAN	

145mya

HERBIVOROUS

**50,000kg
(49 tons)**

**RIODEVA,
TERUEL
PROVINCE,
EASTERN SPAIN**

(TYOOR-ee-ah-SORE-us)
TURIASAURUS RIODEVENSIS

**FIRST HUGE
SAUROPOD
FOUND IN
EUROPE**

This is the first huge sauropod found in Europe – and by huge, think the length of ten cars and the weight of eight elephants. It is by far the biggest land animal ever known to have lived in the continent, and is one of the largest dinosaurs found anywhere in the world. *Turiasaurus* thundered across what's now eastern Spain, living alongside its smaller fellow sauropod *Losillasaurus*, and is known from dozens of bones found near the town of Riodeva. Its teeth were heart-shaped and covered in wrinkled enamel, which helped grind through tough plant matter including leaves, stems and shoots. Its upper forearm bone was the length of a human adult, and it bore a claw on its back feet the size of a rugby ball. The vertebrae found suggest a row of spikes or at least a ridge running down its back. Before *Turiasaurus'* discovery, mega-sauropods were most associated with Africa and America – now we know that these gigantic herbivores also munched their way around western Europe in the Late Jurassic.

**30m
(98ft)**

(MIR-ah-GUY-ah)
MIRAGAIA LONGICOLLUM

JURASSIC

TITHONIAN	
KIMMERIDGIAN	LATE
OXFORDIAN	
CALLOVIAN	
BATHONIAN	MIDDLE
BAJOCIAN	
AALENIAN	
TOARCIAN	
PLIENSBACHIAN	EARLY
SINEMURIAN	
HETTANGIAN	

6m
(20ft)

150–143
mya

HERBIVOROUS

2000kg
(1.9 tons)

This strange stegosaur had a long neck more befitting a small sauropod – in fact its neck contained more vertebrae than most sauropods', with 17 opposed to their typical 12 to 15. It marks the conclusion of the stegosaurs' gradual development of longer necks, the reason for which is unclear: maybe so they could eat higher leaves or for sexual selection, the male with the longest neck appearing most attractive to females. When describing *Miragaia* in 2009 after its partial skeleton was found in Portugal, palaeontologists placed it as a close relation to *Dacentrurus*, a large stegosaur known from fossils found in England. Whereas *Dacentrurus* combined plates and sharp spikes along its back, *Miragaia* seems to have borne plates similar to *Stegosaurus*', only smaller.

MIRAGAIA,
PORTUGAL

JURASSIC

TITHONIAN	
KIMMERIDGIAN	LATE
OXFORDIAN	
CALLOVIAN	
BATHONIAN	MIDDLE
BAJOCIAN	
AALENIAN	
TOARCIAN	
PLIENSBACHIAN	EARLY
SINEMURIAN	
HETTANGIAN	

147mya

C

CARNIVOROUS

perhaps
18,000kg
(17.7 tons)

SVALBARD,
NORWAY

OTHER AMAZING ANIMALS OF THE JURASSIC...

(PLY-oh-SORE-us)
◀··· **PLIOSAURUS FUNKEI**

In the seas of the Middle Jurassic lurked a kind of creature that made *Carcharodontosaurus* look feeble and our modern great white shark appear like a pike in a fishing lake.

Two fragmentary fossils were discovered at Svalbard, a Norwegian island in the Arctic circle, in the middle of the last decade and informally given the memorable monikers 'Predator X' and 'The Monster' – then in late 2012 they were scientifically published as specimens of this hitherto unknown species of pliosaur. *Pliosaurus funkei*'s massive mouth was lined with foot-long spiked triangular teeth, which crunched down on victims such as squid and long-necked plesiosaurs with a biting force probably several times that of *Tyrannosaurus*. It swam using its front two flippers – until it spied its prey, when it also used its rear flippers to provide a burst of speed before attacking.

The fossils were badly fractured after enduring aeons of freezes and thaws, and they degraded further while drying out in the University of Oslo laboratory where Norwegian scientists studied them. Because of the fossils' condition *P. funkei*'s size can only be estimated, but the bigger specimen's skull seems to have been 2m–2.5m (6ft–8ft) long, which along with a few ribs, vertebrae and a huge flipper suggests a body length of 10m–13m (33ft–43ft). Although it turned out to be smaller than suggested in exaggerated reports surrounding the discovery of Predator X and The Monster,

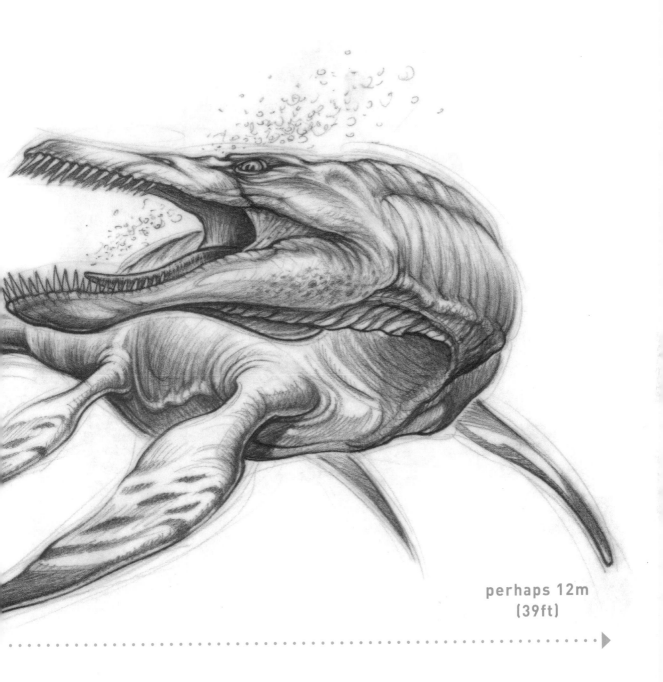

perhaps 12m
(39ft)

it still possessed a power as yet unknown in
the dinosaur world. Put them together with
other unpublished relations such as 'The
Weymouth Bay Pliosaur', found in Dorset, and
'The Monster of Aramberri', from Mexico, and
it's clear that some huge and horrific beasts,
truly the stuff of nightmares, ruled the seas in
Jurassic times.

JURASSIC

TITHONIAN	
KIMMERIDGIAN	LATE
OXFORDIAN	
CALLOVIAN	
BATHONIAN	MIDDLE
BAJOCIAN	
AALENIAN	
TOARCIAN	
PLIENSBACHIAN	EARLY
SINEMURIAN	
HETTANGIAN	

199–189
mya

P

PISCIVOROUS

90kg
(200lb)

BRITAIN

(IK-thee-oh-SORE-us)
ICHTHYOSAURUS COMMUNIS

Other species: I. breviceps, I. conybeari, I. intermedius

With its oversized eyes, protruding snout and fish-like body, this is probably the best known sea creature of the Jurassic period. The English fossil-hunter Mary Anning discovered the first identified specimen on the Dorset coast in 1811 and a decade later William Conybeare and Henry de la Beche named it *Ichthyosaurus*, the 'fish lizard'. Those huge eyes provided detailed vision in the murky depths, helping it pursue its diet of fish and squid.

Hundreds of fossils have been found, most of them in British rocks, and some female specimens have the bones of babies in their wombs, showing that they bore live young rather than laying eggs. Others preserve the outline of *Ichthyosaurus*' body shape, with its distinctive dorsal and tail fins. Despite resembling a fish, it was a reptile descended from creatures that moved from land into water back in the Triassic.

2m
(6ft 6in)

(RAM-foh-RINK-us)

RHAMPHORHYNCHUS MUENSTERI ····▶

JURASSIC

TITHONIAN	
KIMMERIDGIAN	LATE
OXFORDIAN	
CALLOVIAN	
BATHONIAN	MIDDLE
BAJOCIAN	
AALENIAN	
TOARCIAN	
PLIENSBACHIAN	EARLY
SINEMURIAN	
HETTANGIAN	

150mya

P

PISCIVOROUS

4.5kg
(10lb)

GERMANY

**Wingspan
1.8m (6ft)**

The same German limestone that preserved *Archaeopteryx*'s feathery outline has also yielded beautiful fossils of another flying creature from the Jurassic. *Rhamphorhynchus* was a medium-sized pterosaur with a long slender beak containing sharp, needle-like teeth. It would have dashed and swooped across lakes and rivers in the Jurassic, scooping fish and amphibians into its jaws. Its wings spanned a metre and were formed from a membrane that is clear to see in several fossils. Several species have been recognised at times, but now they are considered to be the same one at different stages of development.

JURASSIC

TITHONIAN	
KIMMERIDGIAN	LATE
OXFORDIAN	
CALLOVIAN	
BATHONIAN	MIDDLE
BAJOCIAN	
AALENIAN	
TOARCIAN	
PLIENSBACHIAN	EARLY
SINEMURIAN	
HETTANGIAN	

197–195 mya

C

CARNIVOROUS

2kg (5lb)

DORSET, ENGLAND

(di-MORE-fo-don)

DIMORPHODON MACRONYX

Wingspan 1.2m (4ft)

This huge-headed pterosaur swooped and soared around Europe's coastlines in the Late Jurassic. It had the wingspan of a large seagull and a high-beaked outsize skull like a puffin's, but its two kinds of teeth – fangs at the front, grinding pegs at the back – suggest a diet of insects, small animals and carrion rather than wriggly fish.

On land it was probably a quadruped, as other pterosaurs' fossilised trackways reveal them to have crawled on all fours. Mary Anning discovered the first known British pterosaur fossil in the Blue Lias cliffs near Lyme Regis in 1828 but this fossil was missing the animal's distinctive skull. A year later William Buckland named it *Pterodactylus macronyx* but in 1858, after the discovery of a complete *Dimorphodon*, Richard Owen granted the present genus name while retaining the species name that describes the large claws on its hands. It would have used them for climbing up steep surfaces to heights from which it could leap off and fly.

(terr-oh-DACK-till-us)

PTERODACTYLUS ANTIQUUS

The most famous and earliest-discovered member of the pterosaur family had a crest jutting back from its head and long tapering jaws shaped like a spear-tip that were ideal for catching fish, though it probably also ate small land animals. More than 20 fossil skeletons of varying maturity have been found, many of them in Germany, and the first were described in 1784. Its name means 'wing finger', which describes a characteristic of all pterosaurs. A hugely elongated fourth finger supported a skin membrane that stretched to its hind limbs – think something closer to a bat's wing than a bird's.

**Wingspan
70cm (2ft)**

JURASSIC

TITHONIAN	
KIMMERIDGIAN	LATE
OXFORDIAN	
CALLOVIAN	
BATHONIAN	MIDDLE
BAJOCIAN	
AALENIAN	
TOARCIAN	
PLIENSBACHIAN	EARLY
SINEMURIAN	
HETTANGIAN	

150–148 mya

P

PISCIVOROUS

4.5kg (10lb)

GERMANY

155

WALKING WITH DINOSAURS...

where they left their footprints in North America

The skeletons of dinosaurs – along with many other of their fossils, such as their egg clutches, droppings, and skin impressions – can give palaeontologists a great deal of information about the biology of these extinct animals. But their trackways – the actual footprints they left behind as they moved from one place to another – reveal actions during their lives, such as their walking or running speeds, and occasionally behavioural information about the movements of several individuals in groups.

Dinosaur footprints have been sighted in North America for more than 200 years, but the study of dinosaur footprints didn't begin until the 1860s. Before that time, it was unclear what kinds of animals made the tracks, some palaeontologists thinking they were made by giant birds. With the discovery and study of dinosaur bones in North America in the late 1850s, the tracks could finally be identified as dinosaurian.

Dinosaurs left their footprints in mud along the shores of ponds, lakes, and inland seas where they went to drink or feed. Some trackways look as if meat-eating dinosaurs were following plant eaters, but it is very difficult to tell exactly when different individuals may have made

their tracks. While individual trackways in close proximity may have been made at the same time, they just as easily could have been made on different days or even weeks apart. Tracks can remain in mud for several weeks before being covered with more mud or sand.

The study of animal traces is called ichnology, and since footprints are traces of animals, dinosaur footprints are known as 'ichnofossils'. Here are a few locations in North America where you can see clear proof that they once roamed this land.

Glen Rose, Texas

The most famous dinosaur footprint site in North America is located in Dinosaur Valley State Park along the Paluxy River near Glen Rose, Texas. The footprints were made by sauropod, theropod, and ornithopod dinosaurs about 120 mya as they walked along a muddy tidal flat near a sea. Some sauropod tracks are a meter across, and some trackways are more than 100 feet in length. Some of the sauropod trackways are parallel to one another, suggesting that these dinosaurs were walking together. The dinosaurs that lived in that area at that time – and therefore could have been the animals that left the footprints – include the theropod *Acrocanthosaurus*, the sauropods *Astrophocaudia* and Sauroposeidon, and the ornithopod *Tenontosaurus*.

Connecticut Valley, Connecticut, USA

Discovered in 1802, the first dinosaur tracksite found in North America is located in the Connecticut River Valley, near the town of Rocky Hill in central Connecticut. Most of the tracks are enclosed in a geodesic dome within the confines of Dinosaur State Park. The majority of tracks were made by a theropod dinosaur during the Early Jurassic period about 200 mya. It is thought that the theropod dinosaur may have been similar to *Dilophosaurus*. Some trackways represent prosauropods, the bipedal predecessors of the sauropods. The footprints of the Connecticut Valley are found in sandstone sediments that were deposited along the shores of shallow lakes. At the time these lakes existed, they were located near the center of the supercontinent Pangaea, very near the equator.

Purgatoire River (Picketwire Canyonlands Tracksite), Colorado, USA

The largest dinosaur footprint site in North America is located on public land along the Purgatoire River in southeastern Colorado. More than 1,300 footprints – including more than 100 trackways – have been mapped at the site. The tracks are preserved in limestone that was deposited along the shore of a freshwater lake during the Jurassic Period, about 155 mya. The tracks were made by sauropods, theropods, and ornithopods. The sauropods may have been *Diplodocus* and *Apatosaurus*, the theropod likely *Allosaurus*, and the ornithopod possibly *Camptosaurus*. Numerous trackways are parallel and non-overlapping, suggesting the animals were walking together as a group.

Dinosaur Ridge, Colorado, USA

Dinosaur Ridge is one of the most famous dinosaur sites in North America, known primarily for its Jurassic Period skeletons (from 155 mya) of dinosaurs such as *Stegosaurus*, *Diplodocus*, and *Allosaurus*. These dinosaur skeletons are found on the west side of Dinosaur Ridge. On the east side of Dinosaur Ridge is sediment laid down during the Early Cretaceous period, 120 mya. Here you can see the trackways, mostly of some kind of ornithopod possibly related to *Iguanodon*, and a theropod dinosaur possibly related to *Acrocanthosaurus*. The rock layers in which the footprints are preserved are tilted up at a steep angle so that they are easily viewed from below, where there are interpretive signs for visitors. The area is known as the Morrison Fossil area, a National Natural Landmark.

Red Gulch, Wyoming, USA

The Red Gulch dinosaur trackway is one of the most recently discovered dinosaur trackways in North America. Found in 1997, the site is on public land managed by the Bureau of Land Management – visitors will find interpretive signage to help make sense of the site. There are more than 125 discrete trackways in a space of about 1,600 square meters. A wide variety of dinosaurs are represented, including various sauropods, theropods, and ornithopods. These dinosaurs lived in the area 170 mya, during the Middle Jurassic period. The footprints are preserved in limestone sediments that represent the shoreline of the Sundance Sea, a marine embayment in western North America.

Wasson Bluff, Nova Scotia, Canada

Wasson Bluff, Nova Scotia, has produced the world's smallest dinosaur tracks, representing theropod dinosaurs the size of birds as small as the robin. It is not known if the tiny dinosaurs represent babies or small adults. These little dinosaurs lived 200 mya in a rift valley known as the Fundy Basin. At that time, 200 mya, Africa had just broken free from North America and was moving eastward, opening up a seaway that would eventually become the Atlantic Ocean. These tiny dinosaur tracks can be seen at the Parrsboro Rock and Mineral Museum in Parrsboro, Nova Scotia.

FOOTPRINTS
IN THE SAND

It might seem strange that something so delicate could survive so long. Preserved footprints are actually called 'trace fossils'. These are fossils left while the animal was alive (other examples being coprolites and eggs); true fossils are the remains of the animal once it is dead. The process works like this: a dinosaur leaves a clear footprint in mud or damp sand, which is dried hard by the sun before it can be worn away. A flash flood then sweeps a new layer of sand, grit and gravel across the whole area, filling the imprint. Over a long period both layers are compressed into strata of sedimentary stone. After being hidden from view for aeons, they become revealed again, usually by a cliff fall that sends boulders tumbling on to a beach. For every imprint there is also a stone cast formed by the layer of sand that settled in the dinosaur's footprint, but often the footprint is only revealed by this cast being worn away.

QUIZ ON THE JURASSIC

1. The great landmass known as Pangaea split into two supercontinents in the Early Jurassic, one to the north and the other to the south. What were their names?

2. Why are fossils of sauropods' skulls so seldom found?

3. In 1824 William Buckland made the first scientific description of a dinosaur. What name did he give it?

4. When Richard Owen coined the word 'dinosaur' in 1842, which three genera did he describe?

5. *Australovenator*, *Austrosaurus* and *Ozraptor* were all discovered in which country?

6. What feature made *Spinophorosaurus'* and *Shunosaurus'* tails different from any other known sauropods'?

7. The 'Bone Wars' in late 19th-century America were a feud between which two rival palaeontologists?

8. Why was *Gasosaurus* so-called after it was discovered in China?

9. Palaeontologist John Ostrom's study of which dinosaur caused him to propose an evolutionary link between theropods and modern birds?

10. The sauropod once termed *Brachiosaurus brancai* is now known by which name?

11. How old was Mary Anning when she found her first complete ichthyosaur fossil?

12. Why did *Brontosaurus* become known as *Apatosaurus*?

13. Fossils left when an animal was alive rather than dead – e.g. footprints, eggs and coprolites – are known by what term?

14. Who gave the famous *Diplodocus* skeleton to the Natural History Museum, and in which year?

15. How is the sea creature dubbed 'Predator X' now more properly known?

For answers see page 340.

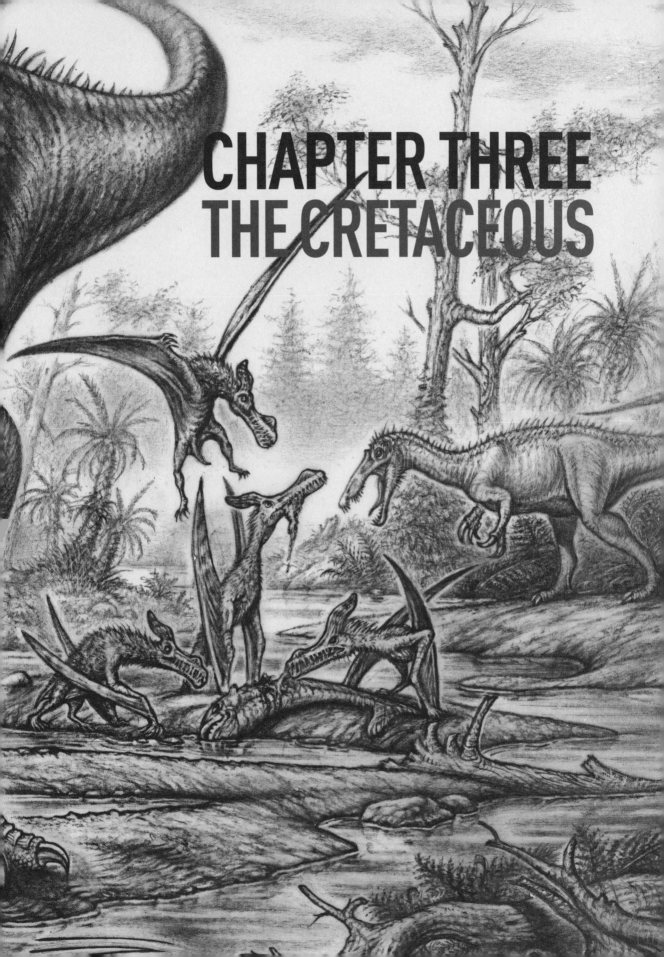

CHAPTER THREE
THE CRETACEOUS

THE CRETACEOUS

No great extinction opens this epoch, only a geological trend that saw more chalk formed than in any other time within the last 500m years, which led a German geologist to name it the Kreidezeit or 'chalk period'. This term was later Latinised into Cretaceous; the limestone-rich Greek island of Crete owes its name to the same derivation. And so while Pangaea dispersed further, with the southern landmass of Gondwana splitting into something approaching the arrangement we recognise from today's atlas, the dinosaurs flourished, diversifying further into some of the most amazing forms to have inhabited Earth. The first spikey-skulled ceratopsians and bone-headed pachycephalosaurs evolved, while monstrous carcharodontosaurs roamed South America alongside the lithe and lethal abelisaurs and the immense herbivorous titanosaurs. In western North America tyrannosaurs became the most advanced meat-eaters known, blending a brain-power and bite-power both unrivalled among fellow land animals of their

time. The warm seas teemed with ammonites, gigantic pterosaurs flapped through the skies, and on land the first flowers began to bloom. Mammals began their ascent, birds became established – and then it all ended with the most almighty bang... except that it didn't end, for we remain surrounded by dinosaurs to this day. Those that survived joined the mammals, fish, reptiles, flowering plants and trees to form the template for our present array of flora and fauna. As the terrestrial dinosaurs' world ended, the one that we recognise was just beginning.

MAASTRICHTIAN	
CAMPANIAN	
SANTONIAN	LATE
CONIACIAN	
TURONIAN	
CENOMANIAN	
ALBIAN	
APTIAN	
BARREMIAN	EARLY
HAUTERIVIAN	
VALANGINIAN	
BERRIASIAN	

140mya

C

CARNIVOROUS

1000kg
(0.9 tons)

EAST SUSSEX,
ENGLAND

(BECK-el-SPY-nacks)

BECKLESPINAX ALTISPINAX

This English carnivore remains poorly understood more than 150 years after Sir Richard Owen first tried to analyse its paltry remains. *Becklespinax* is only known from three tall-spined back vertebrae that for a time were mistaken for *Megalosaurus* vertebrae: this is why the Victorian model of the latter dinosaur at Crystal Palace has a small hump between its shoulders. Suggestions that it was a spinosaur like its English near-contemporary *Baryonyx* have also been brushed off; it's possible but the vertebrae look more like those of an allosauroid. *Becklespinax* remains a mystery but was granted its own genus in 1991, when George Olshevsky named it for Samuel Beckles, the 19th-century fossil-hunter who made the find in Battle, East Sussex. Exactly where in the Hastings Formation rocks he unearthed it is unknown, which makes it hard to date *Becklespinax* precisely.

Beckles was employed by Owen to scour the south coast's fossiliferous rockfaces and return him the results. He left a notable legacy in the Dorset landscape: a huge excavation across 600 square metres (6500sq ft)of the Isle of Purbeck peninsula, now known as Beckles' Pit. Owen gave a small ornithischian discovered by Beckles the name *Echinodon becklesii* in his honour. As for *Becklespinax*, we can be certain it was a medium-sized theropod, so can surmise that it hunted small to medium-sized sauropods – perhaps *Pelorosaurus* or *Xenoposeidon*, though we cannot place its exact contemporaries because of the uncertainty over the location of the fossil.

8m
(26ft)

THE WEALDEN GROUP

In the Early Cretaceous a huge expanse of floodplains and great meandering rivers covered much of southern England, stretching from just south of where London is now, down to northern France. The sediments left behind are now known as the Wealden Group, the name deriving from the Weald area in Kent. These deposits yielded *Becklespinax*'s bones and those of many other English dinosaurs including *Iguanodon* and *Hypsilophodon*.

MAASTRICHTIAN	
CAMPANIAN	
SANTONIAN	
CONIACIAN	LATE
TURONIAN	
CENOMANIAN	
ALBIAN	
APTIAN	
BARREMIAN	
HAUTERIVIAN	EARLY
VALANGINIAN	
BERRIASIAN	

130–120
mya

H

**HERBIVOROUS,
HIGH BROWSER**

25,000kg
(24.6 tons)

**ARAGON,
SPAIN**

(A-rag-oh-SORE-us)

ARAGOSAURUS
ISCHIATICUS

This hefty herbivore roamed what is now the
Aragon region of Spain and may have been
a relative of *Camarasaurus* (page 100), its
American contemporary, though *Aragosaurus*
had longer arm bones so would have stood
a little taller. Like other sauropods it had an
insatiable appetite for vegetation, feeding on
leaves from coniferous trees and rearing up on
to its hind legs to reach the highest branches.
With only fragmentary remains known it is hard
to be precise about *Aragosaurus'* appearance.
Its discovery in 1987 in the Castellar Formation
at La Rioja helped to confirm the worldwide
spread of sauropods in the Early Cretaceous.

**18m
(59ft)**

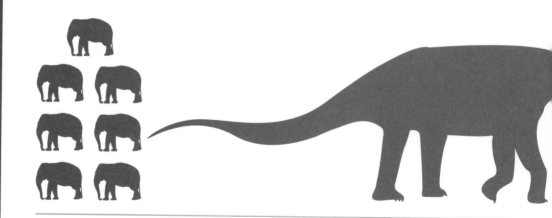

(my)
MEI LONG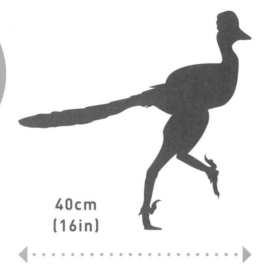

CRETACEOUS

MAASTRICHTIAN	
CAMPANIAN	
SANTONIAN	LATE
CONIACIAN	
TURONIAN	
CENOMANIAN	
ALBIAN	
APTIAN	
BARREMIAN	
HAUTERIVIAN	EARLY
VALANGINIAN	
BERRIASIAN	

MEANING
'SLEEPING
SOUNDLY'

40cm
(16in)

125mya

Its genus name is the Chinese for 'sleeping soundly', which is what the fossilised example discovered in 2004 was doing when it was asphyxiated by volcanic gases and then buried in ash about 140mya. This little duck-sized troodontid's bones were preserved in three dimensions with its snout nestled beneath one of its forelimbs, just as a penguin sleeps with its head tucked behind a wing. As such, it represents a little more evidence of a behavioural link between dinosaurs and birds. *Mei* also has the distinction – shared with *Kol*, an alvarezsaurid found in Mongolia – of having the shortest genus name among the dinosaurs.

C

CARNIVOROUS,
EATING INSECTS
AND SMALL
ANIMALS

0.4kg
(1lb)

LIAONING,
CHINA

169

CRETACEOUS

MAASTRICHTIAN	
CAMPANIAN	
SANTONIAN	LATE
CONIACIAN	
TURONIAN	
CENOMANIAN	
ALBIAN	
APTIAN	
BARREMIAN	EARLY
HAUTERIVIAN	
VALANGINIAN	
BERRIASIAN	

130mya

C

CARNIVOROUS

225kg
(500lb)

ISLE OF WIGHT,
ENGLAND

170

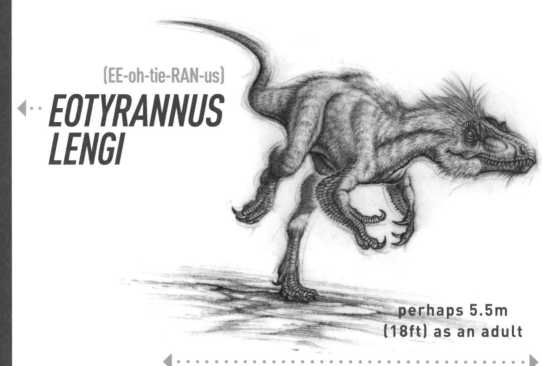

(EE-oh-tie-RAN-us)
EOTYRANNUS LENGI

perhaps 5.5m
(18ft) as an adult

This powerful feathered carnivore's discovery revealed that around 60m years before the famed tyrannosaurs of the Late Cretaceous, a scaled-down model terrorised little herbivores across southern England.

The tyrannosaurs' line began back in the Middle to Late Jurassic with primitive creatures such as *Proceratosaurus* (page 79), *Stokesosaurus*, *Aviatyrannis* (page 110) and *Guanlong* (page 77).

Then in the Early Cretaceous came *Eotyrannus*, whose name means 'dawn tyrant'. Along with the Chinese theropod *Dilong* it forms part of the second phase of the tyrannosauroids.

By this stage they had evolved a heavy skull and upper-jaw front teeth that were D-shaped in cross section – both typical tyrannosaur traits. But unlike *T. rex* and *Tarbosaurus'* tiny two-fingered forelimbs, *Eotyrannus* had long arms and three-fingered hands; in fact its second finger was as long as its forearm. It would have used them when hunting plant-eaters such as the small *Hypsilophodon* and perhaps the larger *Iguanodon*. It is also likely to have been a fast runner, which would have helped it catch prey but also escape its own predators, such as the spinosaur *Baryonyx* and the allosauroid *Neovenator*.

The only known *Eotyrannus* fossil, which is around 40 per cent complete, was found in 1996 in Wessex Formation mudstone high up in the Isle of Wight's cliffs. The lack of fusion between some of the bones suggests that the fossil is that of a sub-adult, so its 4.5m length would have been exceeded by fully grown individuals. Although no feather impressions were preserved, *Dilong* is known to have been feathered so it is most likely that *Eotyrannus* and other early tyrannosaurs were too.

(gas-TONE-ee-ah)
GASTONIA BURGEI

CRETACEOUS

MAASTRICHTIAN	
CAMPANIAN	
SANTONIAN	LATE
CONIACIAN	
TURONIAN	
CENOMANIAN	
ALBIAN	
APTIAN	
BARREMIAN	EARLY
HAUTERIVIAN	
VALANGINIAN	
BERRIASIAN	

Numerous slabs of Utah stone bearing rugged bumps and ridges give a hint of the impenetrable exterior shielding the back of this hefty nodosaurid.

The jumbled remains of around 30 *Gastonia* were found in the same rock formation, along with a fossilised *Utahraptor*, which gives a clue as to why it required such armour. As well as the plating on its back, it bore sharp spikes jutting up from its flanks for added protection. Like other nodosaurids, it differed from ankylosaurids in lacking a tail club.

James Kirkland named *Gastonia burgei* in 1998 in honour of Robert Gaston, who discovered the first bones, and Don Burge, the founder of Utah's Prehistoric Museum.

126mya

H
HERBIVOROUS

1900kg
(1.8 tons)

4–6m
(13–19ft)

UTAH,
USA

MAASTRICHTIAN
CAMPANIAN
SANTONIAN
CONIACIAN
LATE
TURONIAN
CENOMANIAN
ALBIAN
APTIAN
BARREMIAN
HAUTERIVIAN
EARLY
VALANGINIAN
BERRIASIAN

124–122
mya

C

CARNIVOROUS

1.5kg
(3.5lb)

CHINA

(sine-ORN-ith-oh-SORE-us)

SINORNITHOSAURUS MILLENII

90cm
(3ft)

Is this little feathered dromaeosaurid the first confirmed venomous dinosaur? The experts are divided between those who are convinced – in particular the team, led by palaeontologist Empu Gong, who proposed the notion in a 2009 study – those who would like to see more evidence and those who are dismissive. The 'yes' camp cite the presence of supposedly unique grooves in certain long, fang-like teeth that led to cavities in the jaw that could have been venom sacs. But others have responded to Gong's work by saying that the elongated teeth just look that way because they were crushed and squeezed out of the jaw sockets a little during fossilisation, and that other theropods have grooved teeth as well.

Either way, this was a relative of *Deinonychus* and *Utahraptor* and, like these other dromaeosaurs, it had feathers and the characteristic sickle-shaped claw on each foot. But one fact that sets *Sinornithosaurus* apart is that we have an idea of its feathery covering's colour – seemingly a combination of reddish-browns and black – thanks to recent research on fossilised pigment cells (see page 208).

Other species: *S. haoiana*

(nee-OH-ve-nay-tor)

NEOVENATOR SALERII

One of the most thrilling theropods discovered in Britain, this crested carnivore roamed southern England when the area was covered by marshland and populated by grazing herds of *Iguanodon* and *Hypsilophodon*. Their chief threat came in the form of a streamlined predator whose name simply means 'new hunter'. It would also have preyed on ankylosaurs and perhaps even the sauropods with which it shared a habitat. Most interestingly, research in 2010 showed that *Neovenator* was part of a worldwide clade of carcharodontosaurian allosauroids – that is, this English dinosaur was a relative of such lethal beasts as *Australovenator* (see page 226), *Fukuiraptor* and the massive *Megaraptor* (page 247). Together these are now known as neovenatorids. *Neovenator* itself lived in the Early Cretaceous but the last known neovenatorid, a

large Argentinian hunter called *Orkoraptor*, survived to the end of the Cretaceous. This proves that as well as the tyrannosaurs and abelisaurs that are the Late Cretaceous's most familiar predators, there were still some late-surviving allosauroids.

In 1978 the first *Neovenator* bones were seen protruding from the chalk cliffs of the Isle of Wight's southwest coast. It wasn't until 1996 that more were found, and so far about 70 per cent of a single skeleton have been recovered: enough to provide a clear idea of one of Europe's most formidable dinosaurs.

MEANING 'NEW HUNTER'

CRETACEOUS

MAASTRICHTIAN	
CAMPANIAN	
SANTONIAN	LATE
CONIACIAN	
TURONIAN	
CENOMANIAN	
ALBIAN	
APTIAN	
BARREMIAN	EARLY
HAUTERIVIAN	
VALANGINIAN	
BERRIASIAN	

127–121 mya

CARNIVOROUS

1000kg (0.9 tons)

ISLE OF WIGHT, ENGLAND

7.5m (25ft)

173

MAASTRICHTIAN	
CAMPANIAN	
SANTONIAN	LATE
CONIACIAN	
TURONIAN	
CENOMANIAN	
ALBIAN	
APTIAN	
BARREMIAN	EARLY
HAUTERIVIAN	
VALANGINIAN	
BERRIASIAN	

130–125 mya

H

HERBIVOROUS, LOW BROWSER

20kg (45lb)

ISLE OF WIGHT, ENGLAND AND POSSIBLY IN SPAIN

(hip-sih-LOH-foh-don)

HYPSILOPHODON FOXII

For two decades after its discovery in 1849 this hugely abundant little herbivore's bones were thought to belong to *Iguanodon*, but in 1870 Thomas Huxley published the first full description that established it as a new genus. He chose a species name to credit his friend William Fox, who found several skeletons at Brighstone Bay in the Isle of Wight. Like several of Victorian England's leading dinosaur-hunters, Fox was a vicar, but it seems this took second billing to his passion for uncovering the past. Fox's wife once said that it was 'always bones first and the parish second', while Fox himself once wrote to Sir Richard Owen saying 'I cannot leave this place while I have any money left to live on, I take such deep [joy] in hunting for old dragons'.

In *Hypsilophodon* he discovered one of Britain's most widespread dinosaurs of the Early Cretaceous. However, a mistaken interpretation of its toe-bones published in 1882 led to *Hypsilophodon*'s longstanding depiction as a tree-climbing creature, rather like the modern tree-kangaroo, until research in the 1970s showed it to be a ground-dweller. Along with *Iguanodon* it would have browsed foliage in the southern marshes; both would have been hunted by *Neovenator*. *Hypsilophodon*'s only means of defence was its great speed. Its stiff tail served as a counterbalance when running, making it easier to twist and turn and escape predators. Sufficient *Hypsilophodon* fossils have been found together to suggest that they moved in herds and this, along with their size and the fact that their toothy beaks were suited to eating tender shoots, has led them to being dubbed the 'deer of the Cretaceous'.

2.3m (7ft 6in)

MAASTRICHTIAN	
CAMPANIAN	
SANTONIAN	LATE
CONIACIAN	
TURONIAN	
CENOMANIAN	
ALBIAN	
APTIAN	
BARREMIAN	EARLY
HAUTERIVIAN	
VALANGINIAN	
BERRIASIAN	

(a-MARG-a-SORE-us)

AMARGASAURUS CAZAUI

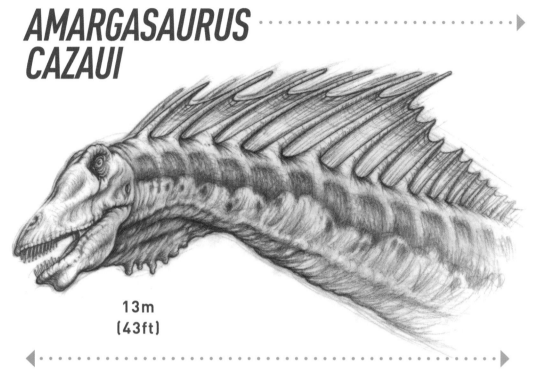

13m (43ft)

130–120 mya

HERBIVOROUS

4000kg (3.9 tons)

LA AMARGA, NEQUEN PROVINCE, ARGENTINA

A tall row of forked spines down its neck and back marks this out as a sauropod unlike any other known. Fellow diplodocoid *Dicraeosaurus* had short forks but *Amargasaurus'* were far more spectacular, the biggest seen in any sauropod. They were highest over the neck and diminished to a single row of small bony stubs at the hips. Some experts think that the skin between the spines formed two parallel sails, while others think this would have rendered its neck inflexible and propose that they were defensive spikes sheathed in keratin. Possible purposes include species identification, temperature regulation and defence, and it is even suggested that *Amargasaurus* could rattle them together to create a warning noise.

This was a small, short-necked diplodocoid, described in 1991 from the discovery in 1984 of a single skeleton in Argentina's La Amarga Formation. The species name refers to Dr Luis Cazau, who prompted the research trip that produced the find.

CRETACEOUS

MAASTRICHTIAN	
CAMPANIAN	
SANTONIAN	LATE
CONIACIAN	
TURONIAN	
CENOMANIAN	
ALBIAN	
APTIAN	
BARREMIAN	EARLY
HAUTERIVIAN	
VALANGINIAN	
BERRIASIAN	

125–121
mya

CARNIVOROUS

1400kg
(1.3 tons)

CHINA

(YOO-tie-RAN-us)
YUTYRANNUS HUALI

If *Eotyrannus* probably bore dino-fuzz, *Yutyrannus* definitely did – and at 9m (30ft) in length it is not only the biggest feathered dinosaur yet found, but the biggest feathered animal ever known to have lived. Its unveiling in 2012 gained worldwide attention. As its name suggests *Yutyrannus* was probably a tyrannosauroid, though it also shares skeletal similarities with the carcharadontosaurs. With three excellent, near-complete skeletons to study, found fossilised together in China's Liaoning province, there is good potential for determining its position in relation to those groups more precisely. The most salient feature was the presence of patches of 15cm-long (6in) plumes sufficient across the three fossils to suggest a full-body covering. These were not like most modern birds' flight feathers or down, but more akin to the plumes on flightless birds such as emus.

Most very large animals are hairless – for instance elephants – but in colder climes hairy varieties also evolve, such as the woolly mammoth. Xu Xing and colleagues suggested the same explanation when describing *Yutyrannus*. As the average temperature in China then was around 10°C according to recent calculations, perhaps its feathers were for insulation in a cool climate.

Aside from its feathers, notable features included a rugged snout comparable to that of fellow feathered tyrannosauroid *Guanlong* (page 77). But the most remarkable implication of *Yutyrannus'* discovery is that it raises the prospect of a wide array of huge coelurosaurs – including the most famous tyrannosaurs – bearing a shaggy covering to some extent.

The three specimens were an adult, a sub-adult and a juvenile, and the adult's skull was around 90cm (3ft) long. *Yutyrannus* had three fingers; it was the advanced tyrannosaurids of the later Cretaceous that had two fingers. It was 40 times larger than the previous biggest known feathered dinosaur, which was *Beipiaosaurus*, a primitive therizinosauroid also from the Early Cretaceous of China. *Yutyrannus'* appearance and no doubt fearsome nature are captured in its name. 'Huali' and 'yu' mean 'beautiful' and 'feathers' in Mandarin, so this is the 'beautiful feathered tyrant'.

**9m
(30ft)**

CRETACEOUS

MAASTRICHTIAN	
CAMPANIAN	
SANTONIAN	LATE
CONIACIAN	
TURONIAN	
CENOMANIAN	
ALBIAN	
APTIAN	
BARREMIAN	EARLY
HAUTERIVIAN	
VALANGINIAN	
BERRIASIAN	

130–125 mya

C

CARNIVOROUS

500kg (1100lb)

UTAH, USA

(YOO-tah-rap-tor)

UTAHRAPTOR OSTROMMAYSORUM

The biggest of the 'raptors' (more properly called the dromaeosaurs) conjures a terrifying image of life in the Early Cretaceous western USA. Imagine a 1.8m (6ft) tall feathered creature running towards you at 20mph wielding sickle-shaped talons up to 23cm (9in) long. That must have been the last thing that many herbivores in its habitat saw, for *Utahraptor* was a devastating predator that probably hunted rather like big cats do today. We can imagine its killing technique from the 'Fighting Dinosaurs' fossil that preserves its smaller relative *Velociraptor* stabbing a toe-claw into a *Protoceratops*' neck – there is no reason to think that *Utahraptor* would have behaved any differently. And a group of fossils of another relative, *Deinonychus*, preserved with the body of a *Tenontosaurus*, suggest that dromaeosaurs perhaps hunted in groups. Some palaeontologists speculate that *Utahraptor* strengthened its kick by propping itself with its stiff tail as it swung its 'killing claw' down into its victim, though others feel its tail could not have been strong enough to use as a support. Others think that dromaeosaurs used these scythe-like claws to pin their prey down, and then they pulled with their jaws to rip out lumps of flesh. Alternatively the terrible gashes it inflicted may have caused its prey to bleed to death, at which point *Utahraptor* returned to devour the carcass.

Jim Jensen and colleagues found its first remains in Utah in 1975 but it was not until the recovery of a large claw in 1991 that they were studied properly. Like other dromaeosaurs, it would have run on two toes with the third bearing this claw aloft from the ground. In 1993 this genus was created with a specific name honouring the eminent American palaeontologists John Ostrom and Chris Mays, who founded Dinamation, a company that created animatronic dinosaurs.

It was almost named *U. spielbergi* instead, as it was hoped that the filmmaker would donate a research fee in return, but no deal could be agreed. However, *Utahraptor* retains one curious link with *Jurassic Park*: although it wasn't known when the film was being made, it was very similar to the devastating '*Velociraptors*' that took a starring role in the film. They were criticised for being far bigger than true *Velociraptor*s – but if you picture them, add feathers, and imagine them a touch larger still, you have a fair idea of *Utahraptor*. And just as *Jurassic Park* was released, *Utahraptor*'s discovery was announced – showing that the idea of a huge raptor wasn't a work of fiction after all.

7m (23ft)

CRETACEOUS

MAASTRICHTIAN	
CAMPANIAN	
SANTONIAN	LATE
CONIACIAN	
TURONIAN	
CENOMANIAN	
ALBIAN	
APTIAN	
BARREMIAN	EARLY
HAUTERIVIAN	
VALANGINIAN	
BERRIASIAN	

125mya

HERBIVOROUS

3200kg
(3.1 tons)

ENGLAND,
BELGIUM AND
GERMANY

180

(ih-GWAAN-oh-don)

IGUANODON BERNISSARTENSIS

A scattering of teeth like an iguana's only far larger emerged from sandstone in Tilgate Forest, Sussex, in 1822 – that much is certain. The details of their discovery are hazy – were they found by Gideon Mantell, his wife or some local quarrymen? – but what matters more are the consequences. These brown chunks of stone provided one of the first glimpses into a lost world. Three years later Mantell, à country doctor and amateur naturalist, illustrated the teeth and described their owner as *Iguanodon*. It was 17 years later that Sir Richard Owen, a brilliant scientist who had become Mantell's adversary, defined the Dinosauria based upon skeletal features shared by this animal, *Megalosaurus* and *Hylaeosaurus*.

Because the earliest *Iguanodon* finds were so minimal, mid-19th-century impressions of its appearance were misguided. Quite understandably Mantell's ideas rested on the assumption it shared an iguana's physique – quadrupedal, low-slung, lumbering – only on

10m
(33ft)

a far bigger scale. An iguana with teeth that size would need to be 18m (59ft) long, so *Iguanodon* was initially considered to be almost twice as large as the moderate-sized herbivore we now see it as. Better remains found in Maidstone, Kent, in 1834 added a spike, vertebrae and limb bones to the gradually developing picture of *Iguanodon*'s skeleton. (You can see the Maidstone specimen in the Natural History Museum.) As is still apparent from the Benjamin Waterhouse Hawkins concrete models built in the early 1850s in the grounds of London's Crystal Palace, early imaginings of its appearance placed a spike upon its nose and a hunch of raised vertebrae upon its neck; the latter feature actually crept in from what turned out to be another dinosaur, *Becklespinax* (page 166). Then in 1878 the image suddenly clarified with a discovery more than 300m (980t) underground in a mine in Bernissart, Belgium.

The miners thought at first they had chipped into petrified wood. It turned out to be the deeply submerged bones of dozens of *Iguanodon*, whose carcasses were washed to a final resting place by a flood around 125m years earlier. The location gave a name, coined by zoologist George Albert Boulenger,

to the one species of *Iguanodon* that experts can still agree on. Like many other long-known dinosaurs, the genus has often served as a 'wastebasket' for superficially similar animals that later turned out to be distinct genera: examples include *Kukufeldia*, *Cumnoria* and, in honour of *Iguanodon*'s finder, *Mantellisaurus*. The true *Iguanodon* was a heavy herbivore that could walk on all fours or bipedally when required. Its arms were three-quarters the length of its legs. Its thumb-spikes may have been a defensive weapon, and perhaps also stabbed open large seeds and fruit; otherwise its diet consisted of tree leaves and low foliage that it cropped off with its sharp beak. The tooth that enabled Mantell to unlock the dinosaurs' era would have been one of more than 100 that were set back in *Iguanodon*'s cheeks and ground up plant matter throughout the day.

Iguanodon left fossils that since Mantell's time have been found across western Europe, and it sits as an evolutionary link between the Early Cretaceous' hypsilophodonts and the duck-billed hadrosaurs. But it is more famous as a link of another kind, one that served to connect modern times with the ancient world of the dinosaurs.

CRETACEOUS

MAASTRICHTIAN	
CAMPANIAN	
SANTONIAN	LATE
CONIACIAN	
TURONIAN	
CENOMANIAN	
ALBIAN	
APTIAN	
BARREMIAN	EARLY
HAUTERIVIAN	
VALANGINIAN	
BERRIASIAN	

125–121 mya

OMNIVOROUS

800g (1.8lb)

JIANCHANG, LIAONING PROVINCE, CHINA

182

(tyan-yoo-LONG)

TIANYULONG CONFUCIUSI

70cm (2ft)

Anyone interested in dinosaurs knows that many theropods bore a downy covering and in some cases true feathers – but the presence of anything similar among ornithischians is a far less familiar notion. *Tianyulong*'s discovery proves this was indisputably true, and a specimen of *Psittacosaurus* shows that it was not an oddity. *Tianyulong* was a small heterodontosaurid whose skeleton, found in the famous Liaoning fossil beds, preserves patches of long bristly filaments on its neck, back and tail. Its publication in 2009 sparked considerable speculation about feathers' origins. It seems that these were more rigid than the soft 'dino-fuzz' associated with coelurosaur theropods. If *Tianyulong*'s were made differently from theropod fuzz, it suggests that the two groups evolved their covering separately. But if they were structurally the same it's reasonable to conclude that bird-hipped and lizard-hipped

dinosaurs inherited the feature from a common ancestor that lived before the split into the two sub-orders – which would take the presence of such proto-feathers all the way back to the dawn of the dinosaurs' age.

Heterodontosaurids such as *Tianyulong* and *Fruitadens* (page 145) gained their name for their mixture of teeth; as well as being equipped to chew vegetation, *Tianyulong* had canine-like fangs that suggest an omnivorous diet. It was an agile little quadruped that scurried around the damp woodlands of China in the Early Cretaceous. Its discovery extended the heterodontosaurids' known geographical range. Previous examples were only known from Africa, Europe and the Americas, making *Tianyulong* (the 'Tianyu dragon', named by Xiao-Ting Zheng) a late survival out on a limb in Asia. The single known specimen was a juvenile so its adult size is uncertain.

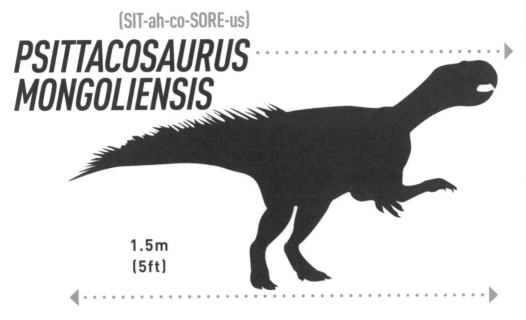

(SIT-ah-co-SORE-us)

PSITTACOSAURUS MONGOLIENSIS

**1.5m
(5ft)**

CRETACEOUS

MAASTRICHTIAN	
CAMPANIAN	
SANTONIAN	LATE
CONIACIAN	
TURONIAN	
CENOMANIAN	
ALBIAN	
APTIAN	
BARREMIAN	EARLY
HAUTERIVIAN	
VALANGINIAN	
BERRIASIAN	

125–100 mya

H

HERBIVOROUS

15kg
(33lb)

CHINA,
MONGOLIA
AND RUSSIA

A parrot-beaked primitive ceratopsian, *Psittacosaurus* was one of the Cretaceous period's greatest success stories: the genus survived around 25m years and comprised more known species than any other non-avian dinosaur. While most left only a single species to the fossil record, *Psittacosaurus* has 11 varieties found in deposits across China, Mongolia and Russia. In 1923 Henry Fairfield Osborn discovered the first in Mongolia and the subsequent discovery of more than 400 specimens marks *Psittacosaurus* out as one of the best-understood dinosaurs. But in the 1990s a fossil illegally exported from China to Germany, and consequently unattributed as yet to any particular species, added new interest to this seemingly familiar herbivore. An exquisite skeleton from the famous Liaoning province shows distinct traces of bristles, comparable to *Tianyulong* but in this case pluming up from the tail. There are around 100 in all, the longest up to 16cm

(6in). The fact that they protrude upwards in a single slender row removes the possibility that they served as insulation. A German team led by palaeontologist Gerald Mayr published the first close study in 2001 and noted that if they were coloured they may have been used for signalling, like modern birds' ornamental feathers. This aside, *Psittacosaurus* was remarkable for its plain appearance when compared with the huge ceratopsians of the Late Cretaceous. Its head was tortoise-like, almost cubic in some especially blunt-snouted species, and the nearest thing it had to *Triceratops'* spikes and frills were a pair of stubby spikes jutting from the back of its jaws. It also differed from its later relatives by being bipedal; research in 2007 showed that its forelimbs could not even touch the ground. Its predators were many but most interestingly included the mammal *Repenomamus* (page 333).

183

AMAZING NORTH AMERICAN DINOSAURS

North America is famous for its dinosaurs, particularly those that come from Alberta, Canada, and the states of Montana, Wyoming, Utah, Colorado, Arizona, and New Mexico. Abundant dinosaur remains come from these areas because of two important geologic events: the uplifting of the Rocky Mountains during the age of dinosaurs, and the erosion of the Rocky Mountains since the extinction of dinosaurs, over the past 65 million years.

The uplifting of the Rocky Mountains produced the geologic setting for the original burial of the dinosaur skeletons, and the erosion of the mountains since that time has exposed the rocks in which the skeletons are now found. Although the Rocky Mountain region is where most of the dinosaur remains are found, there are many other states and provinces that have yielded dinosaur skeletons as well.

In eastern North America, dinosaur remains are rare, but they're occasionally found in Triassic and Early Jurassic rocks along the eastern seaboard, from Nova Scotia to Virginia, and Cretaceous-age dinosaurs have been found in rocks exposed near the Atlantic Ocean, from New Jersey south to Georgia and Alabama. A few dinosaur remains are also known from other states and provinces, such as Texas, Arkansas, Kansas, the Dakotas, California, and Saskatchewan.

While most of the dinosaur remains in the East are Triassic and Early Jurassic, most of the dinosaur remains in western North America are from the Late Jurassic or Cretaceous periods. It's all about the age of the rocks exposed at the ground's surface. Most of the eastern states and provinces are within the area of the Appalachian Mountains, which formed and then began to erode long before the age of dinosaurs – the rocks of an age that would contain dinosaurs are mostly weathered away.

Along the eastern front of the Rocky Mountains in Alberta and Montana, dinosaur-age rocks are exposed at the surface of the ground, and so dinosaur remains are commonly found. The best places to find dinosaurs are where rivers and streams have cut down into the proper age of sediment. Badlands often form in the breaks of the rivers and streams that cut through the Jurassic- and Cretaceous-age rocks, so it is in these badlands that palaeontologists explore for dinosaur remains.

North America is also famous for its dinosaurs because of the spectacular species found within its borders, including *Tyrannosaurus rex*, the most famous of all dinosaurs. *Tyrannosaurus*, *Triceratops*, *Ankylosaurus*, and *Pachycephalosaurus* are some of the last dinosaurs to exist on earth, and their extinction is preserved in the plains of western North America. On pages 185–189 are many of the most notable dinosaurs that once roamed these lands.

1 *Acristavus*, ('non-crested grandfather') lived during the Late Cretaceous Period, 80 million years ago, in Montana and Utah. It was an herbivorous, duck-billed ornithopod dinosaur that reached about 25 feet in length. It is one of the most primitive North American duck-billed dinosaurs known.

2 *Acrocanthosaurus* ('high-spined lizard') lived during the Early Cretaceous Period, 120 million years ago, in Wyoming, Oklahoma, and Texas. It was a large carnivorous theropod that grew to nearly 40 feet in length. *Acrocanthosaurus* had tall spines on its back that were probably for display.

3 *Ankylosaurus* ('fused lizard') lived during the Late Cretaceous Period, 65 to 68 million years ago, in western North America. It was an herbivorous, armoured dinosaur. Adults were about 20 feet long. The armour and club tail are thought to have been used for defence, although some armor on its head may have been for display.

4 *Brachiosaurus* ('arm lizard') lived during the Late Jurassic Period, 153 million years ago, in western Colorado. It was an herbivorous sauropod and is among the largest of dinosaurs reaching 30 feet in height. Unlike *Supersaurus*, which has long hind legs and short front legs, *Brachiosaurus* has long front legs and short hind legs. *Brachiosaurus* likely fed on leaves of very tall tree ferns or conifer trees.

5 *Camptosaurus* ('flexible lizard') lived during the Late Jurassic Period, 150 million years ago, in the western United States. It was an herbivorous bipedal ornithopod that grew to a length of 20 feet. *Camptosaurus* is related to the ancestors of *Iguanodon* and the duck-billed dinosaurs.

6 *Chirostenotes* ('narrow-handed') lived during the Late Cretaceous period, 76 million years ago, in Alberta, Canada, and Montana. It was a toothless oviraptorsaurid theropod whose diet is thought to have been either herbivorous or omnivorous. Oviraptorids have peculiar, short, crested skulls with deep jaws. *Chirostenotes* is thought to have reached a length of 7 feet.

7 *Coelophysis* ('hollow form') lived during the Late Triassic Period, about 200 million years ago, in the southwestern United States. *Coelophysis* was a small carnivorous theropod that reached about 10 feet in length. It has been found in large groups and is thought to have been a very social animal.

8 *Corythosaurus* ('helmet lizard') lived during the Late Cretaceous period, 75 million years ago, in Alberta, Canada. It was an herbivorous, duck-billed ornithopod dinosaur that reached 30 feet in length. The crest on its head is thought to have been used for vocalisation and display. Juvenile *Corythosaurus* did not have a crest.

9 *Deinonychus* ('terrible claw') lived during the Early Cretaceous Period, 110 million years ago, in the western United States. It was a carnivorous theropod, very closely related to *Velociraptor*, that reached a length of 12 feet. Like *Velociraptor* and other dromaeosaurid dinosaurs, *Deinonychus* had a large claw on each hind foot for slicing open its prey.

10 *Diabloceratops* ('devil-horned face') lived during the Late Cretaceous Period, 78 million years ago, in Utah. It was an herbivorous horned dinosaur that reached a length of 18 feet. *Diabloceratops* had large orbital horns and spikes on its shield that were most likely used for display.

11 *Dilophosaurus* ('two-crest lizard') lived during the Early Jurassic Period, 195 million years ago, in Arizona. It was a carnivorous theropod that grew to 23 feet in length. Its elongated snout is suggestive of an animal that fed on fish. The double crest on top of its head was likely for display.

12 *Drinker* (for Edward Drinker Cope) lived during the Late Jurassic Period, 155 million years ago, in Wyoming. It was a small, herbivorous, bipedal ornithopod that grew to 6 feet in length. *Drinker* was related to other small ornithopods, such as *Hysilophodon*.

13 *Dryptosaurus* ('tearing lizard') lived during the Late Cretaceous Period, 67 million years ago, in New Jersey and Delaware. It was a carnivorous, bipedal theropod that grew to about 25 feet in length. *Dryptosaurus* is thought to be a primitive tyrannosauroid.

14 *Edmontonia* ('lizard from Edmonton') lived during the Late Cretaceous Period, 71 million years ago, in western North America. It was an herbivorous armoured dinosaur with large armored spikes on its sides, and no clubbed tail. The body spikes were probably used for display. *Edmontonia* grew to 25 feet in length.

15 *Edmontosaurus* ('Edmonton lizard') lived during the Late Cretaceous period, 65 to 72 million years ago, in western North America. It was an herbivorous, duck-billed dinosaur, up to 48 feet in length. *Edmontosaurus* had a very large, duck-like beak and may have spent a lot of time in water.

16 *Einiosaurus* ('buffalo lizard') lived during the Late Cretaceous Period, 74 million years ago, in Montana. It was an herbivorous, horned dinosaur that reached about 15 feet in length. The nasal horn and spiked shield were most likely for display since the juveniles have short nasal horns and no spikes.

17 *Gryposaurus* ('griffin lizard') lived during the Late Cretaceous Period, 74 to 82 million years ago, in western North America. It was an herbivorous, duck-billed ornithopod that reached about 35 feet in length. *Gryposaurus* is one of the most common duck-billed dinosaurs known.

18 *Hadrosaurus* ('sturdy lizard') lived during the Late Cretaceous Period, 79 million years ago, in New Jersey. It was a duck-billed, herbivorous, ornithopod dinosaur that reached about 25 feet in length. It was one of the first dinosaur skeletons found in North America and the first dinosaur ever to be mounted in an upright pose.

19 *Leptoceratops* ('little-horned face') lived during the Late Cretaceous Period, 66 to 68 million years ago, in western North America. It was a small, herbivorous, protoceratopsian dinosaur related to the larger horned dinosaurs that reached about 7 feet in length. *Leptoceratops* may have been able to switch between walking on all fours and bipedally. Groups of *Leptoceratops* have been found, which suggests that they were social.

20 *Lophorhothon* ('crested nose') lived during the Late Cretaceous Period, 80 million years ago, in Alabama. It was an herbivorous, duck-billed ornithopod that may have reached a length of about 25 feet. The only known skeleton is incomplete and likely represents an immature individual. It is estimated to have been about 13 feet in length. It is the best-known duck-billed dinosaur from the eastern United States.

21 *Maiasaura* ('good mother lizard') lived during the Late Cretaceous Period, about 76 to 77 million years ago, in Montana. It was an herbivorous, duck-billed dinosaur that reached up to 30 feet in length. Nests containing baby *Maiasaura* suggest that the species cared for its young. Hatchling *Maiasaura* were 16 inches long.

22 *Ornithomimus* ('bird mimic') lived during the Late Cretaceous Period, 65 to 72 million years ago, in western North America. Its diet is controversial – it may have eaten both plants and animals, making it an omnivore. Its slender build and long legs suggest it was a very fast runner. It also had a large brain.

187

23 *Oryctodromeus* ('digging runner') lived during the Cretaceous Period, about 95 million years ago, in western Montana and eastern Idaho. It was an herbivorous ornithopod dinosaur, about 7 feet in length. *Oryctodromeus* was a burrowing dinosaur that lived in groups within dens.

24 *Pachycephalosaurus* ('thick-headed lizard') lived during the Late Cretaceous Period, 65 to 68 million years ago, in western North America. It was an herbivorous, bipedal, ornithopod dinosaur, and adults are estimated to have been 15 feet long. The purpose of its dome is controversial: Some scientists think it was for combat, while others think it was for display only.

25 *Pachyrhinosaurus* ('thick-nosed lizard') lived during the Late Cretaceous Period, 70 to 73 million years ago, in Alberta, Canada, and Alaska. It was an herbivorous, horned dinosaur that grew to 26 feet in length. It is unusual in that it has large flattened bosses on its face instead of horns. Its facial bosses and ornamented shield were probably used for display.

26 *Parasaurolophus* ('near-crested lizard') lived during the Late Cretaceous Period, 73 to 76 million years ago, in western North America. It was an herbivorous, duck-billed, ornithopod dinosaur that reached 35 feet in length. The hollow tube-like crest on its head grew to 6 feet in length. It is thought to have been used for vocalisation and display.

27 *Pentaceratops* ('five-horned face') lived during the Late Cretaceous Period, 73 to 75 million years ago, in New Mexico. It was an herbivorous, horned dinosaur that reached about 20 feet in length. The gigantic shield of *Pentaceratops*, as well as its orbital horns, were most likely used for display.

28 *Saurornitholestes* ('lizard bird thief') lived during the Late Cretaceous Period, 72 to 77 million years ago, in Alberta, Canada, and Montana. It was a small, carnivorous theropod that was about 6 feet in length. Similar to its relatives *Deinonychus* and *Velociraptor*, it had a cycle claw on each of its hind feet.

29 *Scutellosaurus* ('little-shielded lizard') lived during the Early Jurassic Period, 196 million years ago, in Arizona. It was a small, herbivorous, bipedal, armoured dinosaur that grew to about 4 feet in length. *Scutellosaurus* was a primitive ancestor of the large armoured dinosaurs, such as *Ankylosaurus*.

30 *Stegosaurus* ('covered lizard') lived during the Late Jurassic Period, 150 million years ago, in the western United States and in Europe. It was an herbivorous, armoured dinosaur that reached 30 feet in length. It likely fed on very short plants, such as ferns. The spikes on its tail were likely for defence, but the plates on its back were more likely for display, although this is controversial.

31 *Styracosaurus* ('spiked lizard') lived during the Late Cretaceous Period, 75 million years ago, in Alberta, Canada. It was an herbivorous, horned dinosaur, reaching about 20 feet in length. The nose horn and spikes on its shield were used for display. Juvenile *Styracosaurs* didn't have shield spikes. *Styracosaurus* remains are found in groups called bone beds, which suggests they traveled in herds.

32 *Supersaurus* ('super lizard') lived during the Late Jurassic Period, 153 million years ago, in Colorado. It was an herbivorous sauropod and is among the largest of known sauropods, reaching 110 feet in length. It may have weighed as much as 35 tons, similar to the weight of an adult gray whale. It is controversial as to whether *Supersaurus* is more closely related to *Diplodocus* or *Apatosaurus*.

33 *Tenontosaurus* ('sinew lizard') lived during the Early Cretaceous Period, about 110 million years ago, in western North America. It was an herbivorous, bipedal ornithopod that reached 25 feet in length. Most of its length was its tail, which consisted of ossified tendons referred to in its Latin name for 'sinew'.

34 *Triceratops* ('three-horned face') lived during the Late Cretaceous Period, 65 to 68 million years ago, in western North America. It was an herbivorous, ceratopsian dinosaur. Adults measured up to 30 feet in length. The horns and shield on *Triceratops* were likely for display, although some scientists think they were for male-to-male combat.

35 *Troodon* ('wounding tooth') lived during the Late Cretaceous Period, 65 to 75 million years ago, in western North America. It was one of the first dinosaurs from North America to be named. *Troodon* was a small, bird-like, carnivorous, theropod dinosaur that reached about 8 feet in length. It had one of the largest brains for its size of any dinosaur.

36 *Tyrannosaurus* ('tyrant lizard') lived during the Late Cretaceous Period, 65 to 68 million years ago, in western North America. It was a large, carnivorous, theropod dinosaur – adults measured up to 40 feet in length. *Tyrannosaurus* had bone-crushing teeth.

MAASTRICHTIAN
CAMPANIAN
SANTONIAN
CONIACIAN
TURONIAN
CENOMANIAN
LATE

ALBIAN
APTIAN
BARREMIAN
HAUTERIVIAN
VALANGINIAN
BERRIASIAN
EARLY

135–120
mya

C

CARNIVOROUS

1200kg
(1.1 tons)

ENGLAND,
SPAIN AND
NIGER

190

(BAR-ee-ON-iks)

BARYONYX WALKERI

First came a single terrifying claw emerging from the 130m-year-old Surrey clay... then a skeleton that conjured the prospect of a nightmarish predator stalking the muddy plains of southern England. *Baryonyx* was a spinosaur, not as immense as its later relative *Spinosaurus* but still a large theropod that, like the rest of its family, primarily ate fish: the presence of scales and bones belonging to the carp-like *Lepidotes* in the fossil's stomach area attests to that. The story of its discovery gives hope to amateur fossil-hunters everywhere. William Walker was a plumber who scoured rock-faces for finds in his spare time, and he was exploring Smokejacks Quarry brickworks at Ockley, near Dorking, when he spied a curious lump protruding from the Wealden Clay. He broke it open to find a claw measuring 25cm (10in) long straight from base to tip, and when he alerted the Natural History Museum in London, their experts went on to procure around 70 per cent of the skeleton from the rocks. The long jaw held 96 teeth, the snout bore a small crest, and the hands had three fingers – one of which bore the 'heavy claw' referred to by *Baryonyx*'s name. This probably served a dual purpose of defence and helping scoop fish from the water into its mouth, just the same as in a grizzly bear today. *Baryonyx*'s discovery helped revolutionise understanding of the anatomy and behaviour of spinosaurs in general, not least its mysterious relative *Spinosaurus* (page 218).

9m
(30ft)

Since then further partial finds have emerged in the Isle of Wight, Spain and Niger, west Africa. The original *Baryonyx* skeleton stayed at the Natural History Museum, where a cast remains on display as one of the most imposing and scientifically significant fossils unearthed in Britain.

CRETACEOUS

MAASTRICHTIAN	
CAMPANIAN	
SANTONIAN	LATE
CONIACIAN	
TURONIAN	
CENOMANIAN	
ALBIAN	
APTIAN	
BARREMIAN	EARLY
HAUTERIVIAN	
VALANGINIAN	
BERRIASIAN	

130mya

C

CARNIVOROUS

3000kg
(2.9 tons)

LAS HOYAS,
CUENCA,
SPAIN

192

(con-CAV-eh-nate-or)
CONCAVENATOR CORCOVATUS

Dinosaur enthusiasts are accustomed to new creatures emerging with ever-stranger adornments – *Guanlong*, the fan-crested hadrosaur *Olorotitan* to name two recent examples – but this theropod seems particularly odd, and it sent the palaeontological world into a spin when its discovery was announced in 2010. For one, *Concavenator* possessed two hugely extended vertebrae just before its hips that must have supported something akin to a rounder version of a shark's dorsal fin. Two, it seems possible that its forearms bore feathers: the ulna is studded with what look like 'quill knobs' (see page 97), the points where feathers' shafts attach to the bone. But *Concavenator* was a primitive carcharodontosaur, a relatively small ancestor of the gargantuan *Giganotosaurus* – not a coelurosaur, the type of dinosaur spanning from tyrannosaurs to maniraptorans whose members are known to have borne feathers.

If this is right, it would change the understanding of feathers' presence among theropods, suggesting an earlier origin and broader prevalence than is presently supposed. However, experts such as British palaeontologist Darren Naish are not convinced: he notes that the knobs on the ulna are unevenly spaced, which may suggest they attached muscles rather than feathers to the bone.

Spanish palaeontologists José Luis Sanz, Francisco Ortega and Fernando Escaso described *Concavenator* after finding its almost complete skeleton at Las Hoyas, a site near the city of Cuenca famed for its beautiful Cretaceous fossils. Its discovery also strengthened the European presence of the carcharodontosaurs, best known for their giant Late Cretaceous species from South America.

6m
(20ft)

(a-gus-TIN-ee-ah)
AGUSTINIA LIGABUEI

15m (49ft)

CRETACEOUS

MAASTRICHTIAN	
CAMPANIAN	
SANTONIAN	LATE
CONIACIAN	
TURONIAN	
CENOMANIAN	
ALBIAN	
APTIAN	
BARREMIAN	
HAUTERIVIAN	EARLY
VALANGINIAN	
BERRIASIAN	

116–100 mya

HERBIVOROUS

8000kg (7.8 tons)

ARGENTINA

Had the plates along *Agustinia*'s back been unearthed on their own, they might have been assigned to an ankylosaur – but the remnants of its limbs seemed to be from a titanosaur. Put together they make for a unique dinosaur: a sauropod with plates and spikes running the length of its backbone, from neck to tail.

This is the most heavily armoured sauropod yet discovered. Aside from nine pieces of armour, the discoverer, Agustin Martinelli, only found a few fragments such as some vertebrae, a hipbone and a 90cm-long (3ft) fibia, the smaller of the two bones in the lower leg.

Jose Bonaparte named it in 1998. Most experts regard *Agustinia* as a titanosaur, though it also seems to share features with the diplodocoids, and a certain definition is unlikely until more remains are discovered.

193

| MAASTRICHTIAN |
| CAMPANIAN |
| SANTONIAN |
| CONIACIAN |
| TURONIAN |
| CENOMANIAN |
| ALBIAN |
| APTIAN |
| BARREMIAN |
| HAUTERIVIAN |
| VALANGINIAN |
| BERRIASIAN |

LATE

EARLY

119–113
mya

H

HERBIVOROUS

300kg
(660lb)

AUSTRALIA

(MIN-mee)

MINMI PARAVERTEBRA

Ankylosaurs were not noted for their intelligence, but even by their standards this primitive example is remarkable for its particularly minuscule brain, which was enclosed within a tortoise-like head. It had long legs – later genera evolved to be massive and proportionally far lower to the ground – and spent its time plodding along and grazing as it went. Unlike many herbivorous dinosaurs, we have a very clear picture of what it ate, thanks to an extremely well-preserved and nearly complete fossil that had remains of a meal in its gut: seeds and fruit swallowed whole, along with fragments of plant material (possibly ferns) that had been nibbled and chopped in its mouth before being ingested. Its first fossils were found at the Minmi Crossing in Queensland, Australia, in the late 1970s. Analysis in 2011 suggested *Minmi* to be the most basal of all known ankylosaurids.

3m
(10ft)

(tie-RAN-oh-tie-tan)

TYRANNOTITAN CHUBUTENSIS

**13m
(43ft)**

MAASTRICHTIAN	
CAMPANIAN	
SANTONIAN	LATE
CONIACIAN	
TURONIAN	
CENOMANIAN	
ALBIAN	
APTIAN	
BARREMIAN	
HAUTERIVIAN	EARLY
VALANGINIAN	
BERRIASIAN	

121–112
mya

C

CARNIVOROUS

7000kg
(6.8 tons)

**CHUBUT
PROVINCE,
PATAGONIA,
ARGENTINA**

Like its fellow carcharodontosaurs, *Tyrannotitan* must have been among the mightiest, most monstrous meat-eaters ever to live. Its prey probably included *Chubutisaurus* – a 23m-long (75ft), 19-tonne (18.7 ton) sauropod, which gives you an idea of just how powerful and voracious this South American predator of the Early Cretaceous must have been.

Little is known about it yet. The only fossil remains found were briefly described in 2005 and given a name that emphasises its probable superior size to *Tyrannosaurus*. The team of researchers led by Argentinian palaeontologist Fernando Novas believe that *Tyrannotitan* was earlier and more primitive than *Carcharodontosaurus* and *Giganotosaurus*.

195

CRETACEOUS

MAASTRICHTIAN	
CAMPANIAN	
SANTONIAN	LATE
CONIACIAN	
TURONIAN	
CENOMANIAN	
ALBIAN	
APTIAN	
BARREMIAN	EARLY
HAUTERIVIAN	
VALANGINIAN	
BERRIASIAN	

122–120 mya

OMNIVOROUS

1.6kg
(3.5lb)

LIAONING,
CHINA

196

(PRO-tark-ee-OP-te-rix)

PROTARCHAEOPTERYX ROBUSTA

PROBABLY ATE INSECTS, SMALL VERTEBRATES AND PLANT MATERIAL

Archaeopteryx is usually credited as the earliest known bird – so you would expect an animal whose name means 'before archaeopteryx' to be a predecessor. Actually *Protarchaeopteryx* lived around 15m years afterwards, and paradoxically appears to have been a more primitive creature. For instance, whereas most palaeontologists think *Archaeopteryx* could fly to some degree (though some research published in 2010, not universally accepted, suggests its feathers may have been too light), *Protarcheopteryx* had feathered limbs but would not have been able to propel itself from the ground: its arms were too short and its feathers too symmetrical. It may have been able to 'parachute', leaping from trees and gliding to

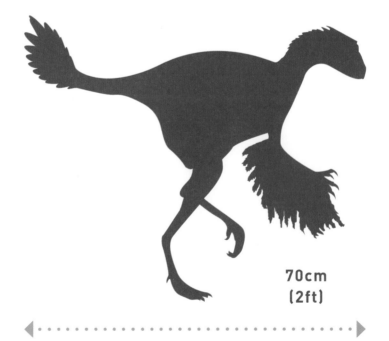

70cm
(2ft)

the ground. It had long legs, a long neck and clawed fingers, suggesting it to have been a fast runner that pursued and grabbed its prey. The feathers probably served as insulation and may have been used in display rituals. In any case, the fact that the fossil was clearly that of a dinosaur – it has features that seem to place it within the Oviraptorosauria family – and yet contained distinct feather impressions is what mattered most to palaeontologists on its discovery in 1996.

This laid to rest the question of whether some dinosaurs were feathered, and moved the debate on to the issue of when feathers became used for flight rather than keeping warm or being used as a display device. In the wake of its discovery the Canadian expert Philip Currie, of the Royal Tyrrell Museum of Palaeontology, said: 'For the first time we have something that is unquestionably a dinosaur with unquestionable feathers. So what we have is a missing link between meat-eating dinosaurs and the earliest bird.'

And that is why this small creature is so very important. As for why it was less advanced than *Archaeopteryx* despite living long afterwards, it is suspected that they shared a common ancestor and evolved separately at different rates.

FIRST CONFIRMED FEATHERED DINOSAUR

CRETACEOUS

MAASTRICHTIAN	
CAMPANIAN	
SANTONIAN	LATE
CONIACIAN	
TURONIAN	
CENOMANIAN	
ALBIAN	
APTIAN	
BARREMIAN	EARLY
HAUTERIVIAN	
VALANGINIAN	
BERRIASIAN	

113mya

C

CARNIVOROUS

uncertain

BENEVENTO
PROVINCE,
ITALY

198

(shippy-ON-ix)

SCIPIONYX SAMNITICUS

The film *Jurassic Park* rekindled enormous popular interest in the world of the dinosaurs – and while it was far from scientifically accurate, it did serve to progress palaeontology in one sense. In the early 1980s an amateur collector named Giovanni Tredesco found the fossil of a juvenile dinosaur while investigating a limestone quarry near the village of Pietraroia in central Italy. He took it home and did little with it for years, until watching *Jurassic Park* spurred him to pass it on to professional palaeontologists. After examining it they declared it the most perfect theropod specimen ever unearthed, retaining such detailed features as muscle fibres, the windpipe, the liver and some remains of the intestines. Had this baby reached maturity, they suspected it would have become a 2m-long (6ft 6in) carnivorous predator. It was also the first dinosaur ever found in Italy, drawn from Cretaceous stone better known for harbouring fine fossils of ancient fish. (The name means 'claw of Scipio', alluding to Scipione Breislak, a geologist who first studied this rock formation, and to a Roman general named Scipio.) Many of the impressions that

palaeontologists have of dinosaurs are informed estimations drawn from a fragment or two of fossilised bone, but with *Scipionyx* the remains are so well preserved that they provided an immediate illustration of this tiny specimen's anatomy. For instance, the liver even retains a pinkish hue, which the experts believe to have been its original colour, and the short length of the intestines showed that it would have digested food relatively quickly. Their contents included remnants of fish and lizard, giving an insight into its diet.

Of course, the fossil did not emerge from the quarry fully exposed and in pristine condition. Palaeontologists had to work for months, using microscopes and tiny chisels, to remove, grain by grain, the layers of sedimentary stone that still cloaked parts of the skeleton. When they were finished the results were astonishing. In 1998, around 113m years after its very brief life ended, this *Scipionyx* made headlines around the world, with its photograph appearing on the cover of *Nature*, the renowned scientific magazine.

24cm (9in)
(adult size estimated
at 2m or 6ft 6in)

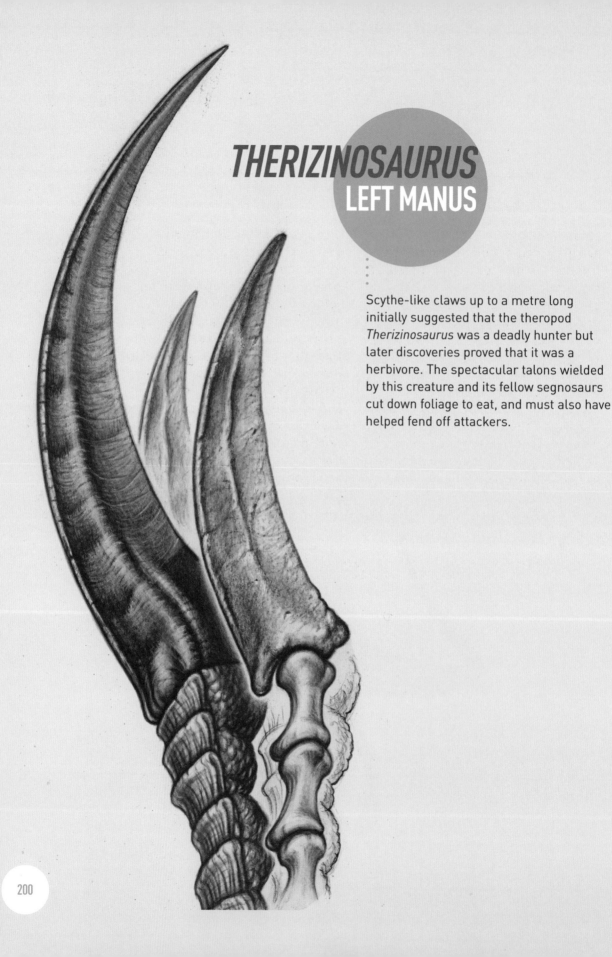

THERIZINOSAURUS
LEFT MANUS

Scythe-like claws up to a metre long initially suggested that the theropod *Therizinosaurus* was a deadly hunter but later discoveries proved that it was a herbivore. The spectacular talons wielded by this creature and its fellow segnosaurs cut down foliage to eat, and must also have helped fend off attackers.

DEINONYCHUS
RIGHT FOOT

Dromaeosaurs such as *Deinonychus*, whose name means 'terrible claw', had an enlarged sickle-shaped talon on each foot. Fossilised trackways show that they ran with it held aloft, poised to swing down when they attacked their prey. *Deinonychus'* 'killing claw' measured around 12cm (5in). In life the fossilised bone that palaeontologists have studied would have been enclosed in a sheath of keratin tapering to a lethally sharp point.

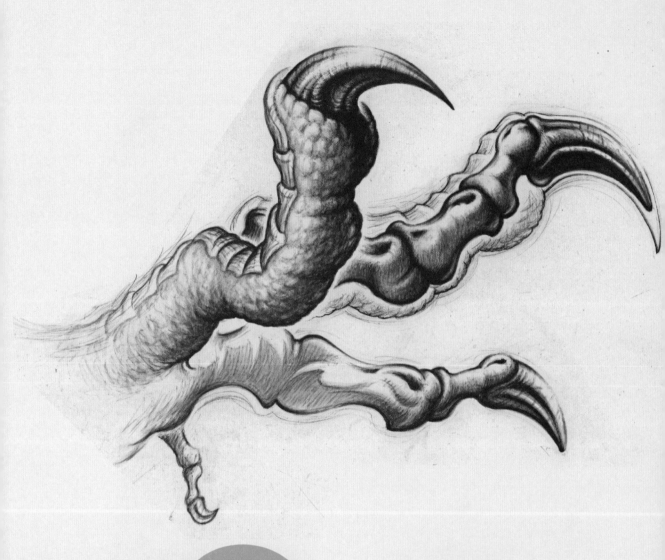

TYRANNOSAURUS
RIGHT FOOT

The famous predator sprinted on three birdlike toes, each bearing a savage hooked claw, while keeping its heel and a smaller fourth digit raised from the ground. This arrangement gave *Tyrannosaurus* extra spring, enabling it to accelerate easily in pursuit of its prey.

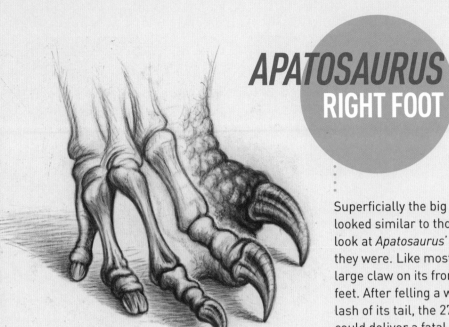

APATOSAURUS
RIGHT FOOT

Superficially the big sauropods' feet may have looked similar to those of elephants but a close look at *Apatosaurus'* bones shows how different they were. Like most sauropods it had a single large claw on its front feet and three on its hind feet. After felling a would-be predator with a lash of its tail, the 27-tonne (26.5 ton) herbivore could deliver a fatal blow by smashing its feet down on the prone theropod. Its feet were also more elongated – measuring around 90cm by 60cm (3ft by 2ft) – and of course far bigger: adult elephants have rounded feet rarely more than 40cm (16in) across.

IGUANODON
LEFT MANUS

Unlike its 19th-century discoverer, Gideon Mantell, we know that *Iguanodon's* spike was part of its hand rather than its nose, but its precise purpose remains uncertain. Possible uses include close-range defence against theropods, ripping into fruit and seed pods, or battling with rivals.

203

CRETACEOUS

MAASTRICHTIAN	
CAMPANIAN	
SANTONIAN	LATE
CONIACIAN	
TURONIAN	
CENOMANIAN	
ALBIAN	
APTIAN	
BARREMIAN	
HAUTERIVIAN	EARLY
VALANGINIAN	
BERRIASIAN	

120mya

C

CARNIVOROUS

4500kg
(4.4 tons)

TEXAS AND
OKLAHOMA,
USA

204

(AK-ro-CANTH-oh-SORE-us)

ACROCANTHOSAURUS ATOKENSIS

11.5m
(38ft)

With a high ridge along its back, a huge pair of hooked extending claws, a metre-long mouth packed with 68 serrated teeth and a body the length of a lorry, this is one of the most spectacular killers of the Cretaceous. It was probably a carcharodontosaur – along with *Mapusaurus*, *Giganotosaurus*, *Tyrannotitan* and the dinosaur that gives that group its name – but its appearance sets it apart from the rest. The ridge's function is uncertain but experts have a good idea of how *Acrocanthosaurus* hunted. A study of its forearms showed that they lacked the flexibility to grab other animals but once it had lunged its head forward and caught its prey in its jaws, it could then impale its victim with its claws, holding the body in place while it ate.

Several sets of large footprints have been found in Texas that seem to show a theropod moving in packs. Those footprints that are now set in stone would have been cast in mud on a coastal floodplain bordering the shallow inland Western Interior Seaway. They are generally attributed to *Acrocanthosaurus*, which would not have been a quick runner – its thighbone was longer than its lower legbones, a sure sign that pace was not among its attributes. But by working in groups this huge hunter could have assailed almost anything it chose to pursue. Its bones were first found in Oklahoma and described in 1950.

(Er-KEH-too)
ERKETU ELLISONI

This sauropod's 7.5m-long (24ft) neck was remarkable for being twice the length of its body. It had the most elongated neck vertebrae and probably the longest neck-to-body ratio of any dinosaur, though not the longest neck per se: that honour goes to *Supersaurus* (page 123), whose neck was double that length. *Erketu* was a relatively small titanosauriform – that is, not actually a titanosaur proper, but part of a broader clade known as somsphospondyls, which included titanosaurs and other relatives such as *Euhelopus* and *Sauroposeidon*. ('Clade' is from the Greek for 'branch'. The word refers to a group comprising a single ancestor and all its descendants – think of it as a single branch on the tree of life and all the twigs shooting off from that branch.) *Erketu* was found in Mongolia's Gobi Desert in 2002, and its name is that of a Mongolian god. When he described it in 2006, Daniel Ksepka of New

York's American Museum of Natural History explained that the largest neck vertebrae had air cavities and a groove running through them that contained a supporting ligament, so *Erketu* didn't have to hold its oversized neck aloft through muscle power alone. Ksepka argued that its posture was probably horizontal rather than diagonal, most naturally holding its neck almost parallel to the ground to cover a wide grazing area rather than straining upwards to reach the treetops. Other palaeontologists, however, believe that holding the neck horizontal would have required huge muscular effort – far more than holding it upright or semi-upright.

LOW
BROWSING

15m
(49ft)

CRETACEOUS

MAASTRICHTIAN	
CAMPANIAN	
SANTONIAN	LATE
CONIACIAN	
TURONIAN	
CENOMANIAN	
ALBIAN	
APTIAN	
BARREMIAN	EARLY
HAUTERIVIAN	
VALANGINIAN	
BERRIASIAN	

120mya

H

HERBIVOROUS,
LOW BROWSER

5000kg
(4.9 tons)

MONGOLIA

CRETACEOUS

MAASTRICHTIAN	
CAMPANIAN	
SANTONIAN	LATE
CONIACIAN	
TURONIAN	
CENOMANIAN	
ALBIAN	
APTIAN	
BARREMIAN	EARLY
HAUTERIVIAN	
VALANGINIAN	
BERRIASIAN	

118–110 mya

H

HERBIVOROUS

60,000kg (59 tons)

OKLAHOMA, USA

206

(SORE-oh-puh-SY-don)

SAUROPOSEIDON PROTELES

When its four neck vertebrae were exposed by rainfall eroding claystone in the grounds of an Oklahoma prison, *Sauroposeidon* was hailed by its finders as the biggest dinosaur of all time – indeed, the biggest creature ever to walk the Earth. This is inaccurate in that *Argentinosaurus* (page 236) was heavier, and *Diplodocus* longer, but *Sauroposeidon* may well have been the tallest. Its huge vertebrae were initially mistaken for fossilised tree trunks, but when they were correctly identified and described they caused a stir: the largest vertebra was 1.4m (4ft 7in) long, suggesting that *Sauroposeidon*'s head was around 20m (65ft) from the ground. Put another way, it stood as high as a six-storey building, or as three very tall giraffes.

Being so tall meant *Sauroposeidon* could eat tough, hard-to-digest leaves at the treetops, beyond the reach of most other dinosaurs. This had great benefits. Food that is hard to digest releases its energy very gradually, which is more efficient than receiving occasional bursts of energy, allowing them a steady source of sustenance.

While its habitat was a vast river delta by the Gulf of Mexico, its name does not imply that it had a marine environment. As well as being the Greek god of the sea, Poseidon had a lesser known role as the god of earthquakes. *Sauroposeidon*'s name derives from the notion that it must have made the ground tremble with its every footstep, though in truth big animals do not really have that effect.

30m (100ft)

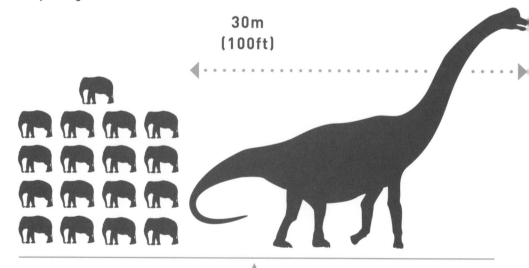

MAASTRICHTIAN	
CAMPANIAN	
SANTONIAN	LATE
CONIACIAN	
TURONIAN	
CENOMANIAN	
ALBIAN	
APTIAN	
BARREMIAN	EARLY
HAUTERIVIAN	
VALANGINIAN	
BERRIASIAN	

(caw-DIP-ter-ix)

CAUDIPTERYX ZOUI

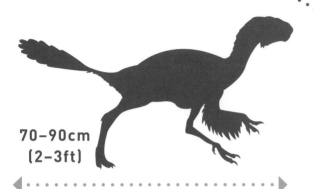

**70–90cm
(2–3ft)**

MEANING
'FEATHERED
TAIL'

120mya

OMNIVOROUS

2.2kg
(5lb)

CHINA

This peacock-sized oviraptorosaur's discovery was a significant moment in the debate over dinosaurs' relationship to birds. A complete fossil found in China's Liaoning province in 1998 has a distinct shadowy outline along the tail culminating in a faint fan of feathers, and the arms also bear similar tufts. These could not have been used for flight, since the arms are too short and the feathers lack the necessary length and stiffness, but in another respect these arm feathers were very like birds', for they had a quill and barbs rather than just being the more typical downy 'dino-fuzz'. So *Caudipteryx* was a dinosaur with recognisable bird feathers. For most palaeontologists this was another piece of evidence for the link between theropods and modern birds.

But if that was the case, where did *Caudipteryx* sit in the evolutionary transition? *Caudipteryx* is considered a basal oviraptorosaur and not a bird. Most palaeontologists place the Oviraptorosauria clade as the sister-taxon to the paravians clade, which comprises deinonychosaurs and birds. (Remember that a clade is a single ancestor and all its descendants. Sister taxa are two branches of this family tree leading off from a shared splitting point. So oviraptorosaurs and paravians are sister taxa within the larger maniraptoran clade.)

Although it couldn't fly and must have used its feathers for display, they had another interesting potential purpose. *Caudipteryx* was primarily a ground-dweller but could probably run up trees – not by clinging on with all four limbs, but by flapping its feathered arms back and forth, which created a suction force that stopped them falling from the trunk as their legs propelled them upwards. This is known as 'wing-assisted incline running'. In 2002 American scientist Kenneth Dial studied this behaviour in galliform birds – chickens, turkeys, pheasants and grouse – and suggested that *Caudipteryx* could have done the same.

RESEARCH INTO DINOSAURS' COLOURS

More than 150 years on from their discovery, our picture of the dinosaurs' world is finally gaining some colour. For years, palaeo-artists have speculated when choosing which paints to choose from their palettes, with most conservatively restricting themselves to dull shades of green, brown and grey – but now, for a few genera at least, a new spectrum is emerging.

In January 2010 a team of British and Chinese researchers announced that they had been studying melanosomes, vessels of pigment so small that 100 of them fit across a human hair, in the fossils of bird-like dinosaurs. They did so by using a scanning electron microscope, which beams electrons at an object to produce a 3-D image of up to 500,000-times magnification.

One creature they studied was *Sinosauropteryx*, a turkey-sized carnivore found in China's Liaoning province, which is known for its finely preserved 'dino-bird' fossils from the Early Cretaceous. Looking at the fossilised remains of proto-feathers, the bristly precursor to feathers sometimes dubbed 'dino-fuzz', they detected melanosomes just like those in modern birds' feathers and mammal hair, including the hair on your head.

These come in different varieties: there are eumelanosomes, which are sausage-shaped and produce black and grey shades, and phaeomelanosomes, which are spherical and produce reddish-brown to yellow colours. In analysing the 'dino-fuzz' preserved on *Sinosauropteryx*'s tail, the researchers found bands that were 'absolutely packed with phaeomelanosomes', in the words of the study's co-author, Professor Mike Benton of Bristol University. This led them to conclude that it had 'chestnut to rufous' coloured stripes. The discovery strengthens the notion that feathers originally evolved for display purposes, only much later developing in a way that enabled birds to fly: perhaps *Sinosauropteryx* used its striped tail when courting a mate or warning off a rival.

The team also studied the feathers of a primitive bird, *Confuciusornis*, from the same fossil site and concluded that it was covered in patches of black, white and brown. This technique had actually been pioneered in 2008 by a team comprising scientists from Beijing in China and Yale University in the USA. Led by Danish researcher Jakob Vinther, they examined a Cretaceous-period bird feather from Brazil and another from Germany that dated back 47m years to the middle of the Eocene epoch. The former bore bars of black and white, the latter was dark with an iridescent sheen, like a starling. And just after the British-Chinese team released its findings, Vinther's team at Yale University announced the outcome of a more in-depth study concerning the 155-million-year-old *Anchiornis huxleyi*, a feathered bird-like dinosaur that predates *Archaeopteryx* by 10m years. Their analysis of 29 samples from all over a complete *Anchiornis* fossil provided the first full-body description of a dinosaur's colours and patterns. This small troodontid resembled a cross between a woodpecker and a pheasant with black and white barred wings and a reddish crown of plumage on its head.

Then in 2011 another team led by American scientist Ryan Carney used a combination of scanning electron microscopy and X-ray analysis to study an *Archaeopteryx* feather. They found it rich in melanosomes, and comparing them with samples from 87 different modern birds showed that this feather was very dark, perhaps black. Whether *Archaeopteryx* was dark all over remains to be seen, but Carney observed that dark pigment cells have a tougher structure than light ones, thus making the feathers stronger – so colour wasn't just about aesthetics, it could also confer a physical advantage.

You'll notice that all the colours mentioned so far remain from a limited range: black, grey, white, brown, rufous red. Of course other types of pigment cells exist – for instance carotenoids, which produce orange, pink and vivid red tones, but unfortunately these structures do not fossilise so well. However they do leave chemical traces that scientists hope to analyse in future. Scientists would not expect to find those pigments in *Sinosauropteryx*, *Anchiornis*, *Confuciusornis* or *Archaeopteryx* in any case, because those four animals were all carnivores – and creatures with pink, orange and red colours gain the necessary carotenoids

through a diet of plants and crustaceans. For instance, flamingos get their pink hue from carotenoids in the algae and brine shrimps they eat. Give them a different diet and they turn white.

However, a new avenue to explore emerged in 2011 with another astonishing discovery: remnants of what appear to be proto-feathers in lumps of amber from Alberta, Canada, preserved for 79m years after a dinosaur briefly brushed against the sticky resin of a tree in a steamy coastal forest. There is no indication yet of their colour or the genus of dinosaur from which they originate, but this find – made when a PhD student re-examined thousands of amber pieces held in storage at the Royal Tyrell Museum – is exciting as it preserves the original biological matter, rather than a version that has been fossilised.

So already a vivid picture is emerging of some feathered dinosaurs' true appearance, though sadly the chances of confirming scaly dinosaurs' colours are very small, as fossilised scales do not appear to retain pigment. But until recently it was assumed that no dinosaurs' colours would ever be known, so who would dismiss the prospect of another revelatory breakthrough? In future artists may find themselves armed with the knowledge to deploy a rainbow palette that colours every aspect of the dinosaurs' world.

MAASTRICHTIAN

CAMPANIAN

SANTONIAN

CONIACIAN

TURONIAN

CENOMANIAN

ALBIAN

APTIAN

BARREMIAN

HAUTERIVIAN

VALANGINIAN

BERRIASIAN

LATE

EARLY

106mya

HERBIVOROUS

90kg
(200lb)

**VICTORIA,
AUSTRALIA**

210

(LEE-ell-IN-a-SAW-ra)

LEAELLYNASAURA AMICAGRAPHICA

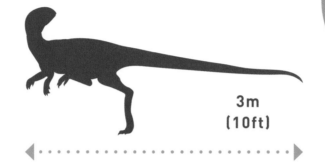

**AS LONG
AS A BOA
CONSTRICTOR**

**3m
(10ft)**

Australian palaeontologists Tom Rich and Patricia Vickers-Rich discovered *Leaellynasaura* at Dinosaur Cove in the southern state of Victoria, which was within the Antarctic Circle during the Middle Cretaceous. While the South Pole was warmer then than now, it still would have been cold and dark for many months of the year, and this ornithopod's discovery was the first to prove that dinosaurs could exist in such conditions. Its skull had notably large eye sockets, which suggest that it had big eyes capable of seeing in the prolonged sunless period of the polar winter. *Leaellynasaura*'s other distinctive feature was its very long tail, three times its body length – the longest of any ornithischian dinosaur. The finders named it in 1989 for their daughter, Leaellyn.

(LOOR-duh-SORE-us)
LURDUSAURUS ARENATUS

· ▶

CRETACEOUS

| MAASTRICHTIAN |
| CAMPANIAN |
| SANTONIAN |
| CONIACIAN |
| TURONIAN |
| CENOMANIAN |
| ALBIAN |
| APTIAN |
| BARREMIAN |
| HAUTERIVIAN |
| VALANGINIAN |
| BERRIASIAN |

LATE

EARLY

This was an oddity: a huge iguanodontian that had the squat bulky form of a sauropod, and may have lived a semi-aquatic lifestyle. North Africa's Tenere Desert was at that time a coastal river delta and *Lurdusaurus'* splayed footbones suggest it was adapted for moving through water. Like other iguanodonians its hands had a thumb-spike – a particularly lethal one, no doubt useful for self-defence given that its low centre of gravity and stocky legs rendered it a very slow mover.

The French palaeontologist Philippe Taquet found the partial remains in 1965 during a research trip to the desert in Niger. It was not until 1988 that a French PhD student granted the remains a name, *Gravisaurus tenerensis*, but while this was used informally for a time, it was technically invalid because the dissertation did not feature a proper description. So in 1999 Taquet, working with American palaeontologist Dale Russell, finally published a name that has much the same meaning: 'heavy lizard from the sands'.

112–99
mya

HERBIVOROUS, LOW AND MIDDLE BROWSER

2500kg
(2.4 tons)

◀ · ▶

**7m
(23ft)**

NIGER, NORTH AFRICA

MAASTRICHTIAN	
CAMPANIAN	
SANTONIAN	LATE
CONIACIAN	
TURONIAN	
CENOMANIAN	
ALBIAN	
APTIAN	
BARREMIAN	EARLY
HAUTERIVIAN	
VALANGINIAN	
BERRIASIAN	

121–112 mya

CARNIVOROUS

5000kg
(4.9 tons)

NIGER, AFRICA

212

(soo-koh-MIME-us)

SUCHOMIMUS TENERENSIS

MEANING 'CROCODILE-LIKE'

11m
(36ft)

Its name means 'crocodile-like', which is a good description of its long slender snout, but *Suchomimus* was almost twice the size of even the hugest crocodiles living today. Like other spinosaurs – and the modern Indian gharial that it particularly resembles – it probably subsisted on fish, gulping them from the lakes and rivers of swampy Cretaceous Africa into a mouth crammed with 100 narrow teeth. The correct term for this diet is 'piscivorous', but it is suspected that *Suchomimus* also scavenged land animals so it is safe to term it a carnivore. It had long, powerful arms so may have dipped them into the water to pluck out its prey. Its more famous relative *Spinosaurus* bore a huge sail on its back and *Suchomimus*' skeleton, found in the Tenere Desert of Niger in 1998, suggests a smaller ridge along the spine, much like the English spinosaur *Baryonyx*. In fact it bears such similarities to that genus that some palaeontologists consider *Suchomimus* an African species of *Baryonyx*. It was larger than the English specimen, however, and may have been a juvenile, in which case it would have grown even bigger – perhaps rivalling *Spinosaurus*.

CRETACEOUS

MAASTRICHTIAN	
CAMPANIAN	
SANTONIAN	LATE
CONIACIAN	
TURONIAN	
CENOMANIAN	
ALBIAN	
APTIAN	
BARREMIAN	
HAUTERIVIAN	EARLY
VALANGINIAN	
BERRIASIAN	

(die-NON-ih-cus)

DEINONYCHUS ANTIRRHOPUS

One of the most terrifying predators of the American Early Cretaceous, *Deinonychus* is responsible for stimulating the 'dinosaur renaissance', the resurgence of interest after a long lull during the mid-20th century.

Or more accurately the late American palaeontologist John Ostrom was responsible, for it was he who studied this large dromaeosaurid's skeleton and in 1969 described an active hunter sporting a scything claw on each foot. His findings vivified our picture of the Mesozoic world: after decades in which dinosaur palaeontology lay largely dormant, bedevilled by the notion that dinosaurs were dull and sluggish reptiles with nothing to teach us about the modern world, Ostrom's observations conjured thrilling new images of warm-blooded, active and intelligent creatures directly linked with the present day. *Deinonychus* was named for the 'terrible claw' on its feet – but while its hands were less spectacular, they were of immense scientific interest. Ostrom observed the bones' extreme similarity

to those of birds' wings and thus revived the old notion of an evolutionary connection between birds and dinosaurs.

Picture the '*Velociraptors*' in *Jurassic Park*: they were actually *Deinonychus*, though the real animal probably had feathers rather than scales and wasn't quite so large. But the filmmakers were right to show it as a deadly hunter. Its habitats included the swampy conditions of western North America where it probably preyed on *Tenontosaurus*, as *Deinonychus* teeth have been found fossilised with the bones of that large ornithopod. It seems the teeth were shed as it bit and gnawed at the herbivore's carcass. To kill such a creature more than twice its length it may have hunted in packs; trackways attributed to *Deinonychus* seem to show several animals running together.

110mya

C
CARNIVOROUS

73kg
(160lb)

MONTANA AND OKLAHOMA, USA

3.5m
(11ft 6in)

| MAASTRICHTIAN |
| CAMPANIAN |
| SANTONIAN | LATE |
| CONIACIAN |
| TURONIAN |
| CENOMANIAN |
| ALBIAN |
| APTIAN |
| BARREMIAN | EARLY |
| HAUTERIVIAN |
| VALANGINIAN |
| BERRIASIAN |

119–99
mya

H

HERBIVOROUS

2000kg
(1.9 tons)

GADOUFAOUA,
NIGER, AFRICA

214

(NEE-jer-SORE-us)
NIGERSAURUS TAQUETI

Why would any dinosaur evolve a mouth shaped like the flat nozzle on a vacuum cleaner? Evolutionary theory dictates that there is a purpose to every animal's form – or at least that there once was, given that many creatures' bodies retain vestiges of structures that have evolved away: think of snakes' leg-spurs, for instance. The evolutionary advantage of *Nigersaurus'* facial features continues to intrigue scientists. Its head was an extraordinary adaptation, the bones so light as to be almost translucent, the flat front of the jaws lined with three rows comprising over 100 tiny sharp teeth – and up to eight replacements stacked beneath each one, ready for when the tooth wore out. In its discoverer Paul Sereno's view it replaced its teeth every month.

In the Middle Cretaceous, *Nigersaurus'* habitat in north-western Africa was damp and swampy. Sereno found the fossil in a region of Niger's Tenere Desert so dry and inhospitable that local people call it Gadoufaoua, 'the place where camels fear to tread'. In 2007 his team digitally scanned the skull and for the first time determined the positioning of a sauropod's brain, inner ear and olfactory bulbs, which control the sense of smell. Doing so granted a good understanding of *Nigersaurus'* (and by extension perhaps other rebbachisaurid diplodocoids') natural posture, suggesting that this relatively short-necked herbivore habitually held its head down to the ground – which therefore implies that it really did work like a vacuum cleaner, the mouth methodically moving from one side to the other as it clipped its way through ferns and other low foliage.

9m
(30ft)

(MIKE-row-rap-tor)
MICRORAPTOR ZHAOIANUS

CRETACEOUS

MAASTRICHTIAN	
CAMPANIAN	
SANTONIAN	LATE
CONIACIAN	
TURONIAN	
CENOMANIAN	
ALBIAN	
APTIAN	
BARREMIAN	EARLY
HAUTERIVIAN	
VALANGINIAN	
BERRIASIAN	

Other species: *M. gui*

90cm
(3ft)

128–124 mya

C

CARNIVOROUS

1kg
(2.2lb)

LIAONING PROVINCE, CHINA

Was this a four-winged dinosaur? Proto-feathers covered this intriguing little dromaeosaur's body, but its front and hind limbs all bore true flight feathers. One view is that it adopted a biplane-like posture, gliding in a shallow U-shaped trajectory between trees in the Asian woodlands, while others think it could fly. Some experts see *Microraptor*'s four wings as an evolutionary dead-end, but others note that eagles also have flight feathers on their legs, suggesting they may have evolved from four-winged ancestors.

Xu Xing described *Microraptor* in 2003 and now an estimated 300 specimens are held in museums and institutions around the world. Many preserve details of feathers, which hint at dark and light bars of colour. Its great abundance in the fossil record shows that *Microraptor* must have been one of the commonest little predators in its habitat. In 2011 Xu and two colleagues described a specimen with the fossilised bones of a small bird in its stomach; another contains remnants of a small mammal's bones. The bird Xu mentioned lived in the trees rather than on the ground, so the fossil provided possible evidence that *Microraptor* hunted in trees. Its claws would have enabled it to scale trunks and branches in pursuit of prey.

This behaviour reinforces the idea of a 'down from the trees' rather than 'up from the ground' origin of flight. *Microraptor*'s eyes seem suited for nocturnal activity, suggesting it hunted under cover of darkness.

8m
(26ft)

(IRR-it-ate-or)
IRRITATOR CHALLENGERI

CRETACEOUS

MAASTRICHTIAN	
CAMPANIAN	
SANTONIAN	LATE
CONIACIAN	
TURONIAN	
CENOMANIAN	
ALBIAN	
APTIAN	
BARREMIAN	EARLY
HAUTERIVIAN	
VALANGINIAN	
BERRIASIAN	

110–100 mya

CARNIVOROUS – FISH, PTERODACTYLS, CARRION

1000kg (0.9 tons)

EASTERN BRAZIL

Few things are more annoying for palaeontologists than studying a strange new fossil and then realising that it has been tampered with. Worse still is when some unscrupulous person has meddled with it in an attempt to increase its value. When British palaeontologist David Martill and his team began working with an 80cm-long (31in) skull found in eastern Brazil, they realised that the illegal fossil-poachers who uncovered it had elongated its snout with plaster to make it look more complete before selling it as a supposed pterosaur fossil. Returning the skull to its natural condition took a great deal of time and effort, and realising they had been duped was extremely irritating. Once restored to its unadulterated state it became apparent that it was a theropod – initially identified as a troodontid-like coelurosaur, but later as a variety of spinosaur, a smaller relative of the enormous *Spinosaurus*. This was still a large creature, mind, which would have subsisted on a similar diet of fish and smaller land animals. In 2004 researchers working with the fossil of a genuine pterosaur found what they believe to be an *Irritator* tooth embedded in a neck bone. The single species is named for Professor Challenger in Sir Arthur Conan Doyle's novel *The Lost World*, a character who leads an expedition to South America to reveal a land of dinosaurs.

217

CRETACEOUS

MAASTRICHTIAN	
CAMPANIAN	
SANTONIAN	LATE
CONIACIAN	
TURONIAN	
CENOMANIAN	
ALBIAN	
APTIAN	
BARREMIAN	EARLY
HAUTERIVIAN	
VALANGINIAN	
BERRIASIAN	

112–97
mya

C

CARNIVOROUS

9000kg
(8.8 tons)

BAHARIYA
OASIS, EGYPT

218

(SPY-no-SORE-us)

SPINOSAURUS AEGYPTIACUS

Foot-long claws on arms strong enough to puncture steel. Crocodilian jaws lined with conical teeth. A dorsal sail the height of a human, atop a body as long as a lorry. Everything about *Spinosaurus* induces awe and fear, for this incredible theropod is the biggest carnivorous dinosaur known to have lived. And while most of the huge hunters represented variations on fundamentally the same body-plan, *Spinosaurus* and its relatives were something different – something extremely specialised and all the more memorable for it. But only now is *Spinosaurus* coming to the forefront as one of the mightiest beasts of the Mesozoic era.

It was discovered in 1912, ten years after *Tyrannosaurus*, but for decades our picture of its appearance was too hazy for *Spinosaurus* to imprint itself on the popular imagination in the same way. A German expedition led by the aristocrat Ernst Freiherr Stromer von Reichenbach brought home from Egypt a series of vertebrae bearing spines up to 1.65m (5ft 5in) high and the lower section of a long, crocodilian-like jaw. These intriguing bones were described, illustrated and housed in the Bavarian State Collection of Palaeontology in the region's capital Munich, but in April 1944 Allied bombs rained down on the area, targeting the nearby Nazi

Party headquarters and devastating the museum. Having spent 95m years preserved in the rocks of the Sahara desert, the *Spinosaurus* fossil was destroyed after three decades' contact with humanity.

Then almost 40 years later a *Baryonyx* (page 190) claw appeared in an English quarry and unlocked *Spinosaurus'* secrets. The fossil subsequently recovered was the first good spinosaur skeleton, looking similar to Stromer's illustrations of those lost *Spinosaurus* remains, only smaller and lacking such extended spines. By comparing the two, palaeontologists deduced *Spinosaurus'* overall proportions, scaling up from the more complete *Baryonyx*. The creature that emerged from their calculations had a body up to 18m (59ft) long, with 2m-long (6ft 6in) arms wielding 38cm (15in) meat-hook claws, and a probable 1.75m (5ft 9in) skull, its elongated jaws studded with curved, conical, interlocking fangs.

So why did it evolve such a strange and specialised form? Doing so allowed the spinosaurs to carve out their own niche alongside more conventional theropods. *Spinosaurus* lived alongside the mighty *Carcharodontosaurus*, which was smaller but probably more powerful and had heftier jaws. Rather than compete

18m
(59ft)

with *Carcharodontosaurus* for dominion of the land, *Spinosaurus* focused on ruling the waters. *Carcharodontosaurus* could feast on sauropod flesh; *Spinosaurus* would gorge itself on fish and shore-dwelling animals, much as crocodiles do today. And if it seems unlikely that something so gargantuan could subsist primarily on fish, the lakes of Cretaceous Africa contained 3m-long (10ft) coelacanths (see page 240) called *Mawsonia*, huge lungfish, and a formidable 8m (26ft) sawfish called *Onchopristis*, among other substantial species. In 1975 fossil fragments found in the red sandstone of Morocco's Kem Kem Desert added more skull material; wedged between a section of jaw and tooth was a probable *Onchopristis* vertebra. The discovery of the well-preserved *Suchomimus* (page 212) revealed that the family's skulls were even narrower than previously expected: less like a crocodile's than a gharial's, with its slender snout leading to a bulbous tip that helps secure wriggling prey in the jaws. This strengthened their ability to catch fish and reduced the likelihood of their killing large terrestrial dinosaurs. Along with its smaller relative *Irritator* (page 217), *Spinosaurus* had nostrils high up its snout, allowing it to breathe while largely immersed in water. A French study in 2009 assessed spinosaur fossils' level of oxygen isotopes (which are higher in aquatic animals) and concluded that they had a semi-aquatic lifestyle, spending as much time in water as today's hippopotamuses and crocodilians.

And its adaptation for fishing may even explain the huge sail along its back, according to an Italian palaeontologist. Cristiano Dal Sasso observed that some herons use their wings to cast shadows over water. Fish quickly seek the shade, at which point the heron plucks them from the water with its beak. Did the vast rounded shadow of *Spinosaurus'* sail perform the same function? Corroborating this is a 2005 study in which Dal Sasso found evidence of sensory points within the long snout that served as motion detectors, enabling *Spinosaurus* to sense fishes' movement underwater while its eyes were above the surface.

But even if they were primarily piscivorous, spinosaurs undoubtedly ate whatever they could. For instance, *Baryonyx* had remnants of a young *Iguanodon* in its gut, though these may have been scavenged.

As with other specialised dinosaurs, *Spinosaurus'* extreme power in one environment proved to be its downfall in another. Around 97mya the North African climate cooled sharply reducing its food supply, and *Spinosaurus* could not adapt. It disappeared from history, and so far North Africa's rocks have yielded only scanty remains just hinting at the size and strangeness of the hugest known theropod dinosaur.

MAASTRICHTIAN	
CAMPANIAN	
SANTONIAN	LATE
CONIACIAN	
TURONIAN	
CENOMANIAN	
ALBIAN	
APTIAN	
BARREMIAN	
HAUTERIVIAN	EARLY
VALANGINIAN	
BERRIASIAN	

(mutta-burra-SORE-us)

MUTTABURRASAURUS LANGDONI

This bulky herbivore is one of the most complete dinosaurs unearthed in Australia despite the fact that after its skeleton's discovery at Muttaburra in 1963, it was first kicked around by grazing cattle, then disturbed by local people who took bones home as souvenirs. Once its significance emerged most were returned, allowing researchers to build a detailed picture of a large iguanodontian with a hollow bump in front of its eyes, which may have heightened its sense of smell or enabled it to make a noise. Its long arms suggest that it could walk on all fours as well as bipedally. Its teeth sliced through tough plants such as cycads, rather than grinding vegetation as others did. Curiously enough, in another *Muttaburrasaurus* find at Lightning Ridge in New South Wales, the teeth have been transformed over time into opal, the semi-precious stone.

100–98
mya

HERBIVOROUS

2800kg
(2.7 tons)

AUSTRALIA

8m
(26ft)

CRETACEOUS

MAASTRICHTIAN	
CAMPANIAN	
SANTONIAN	LATE
CONIACIAN	
TURONIAN	
CENOMANIAN	
ALBIAN	
APTIAN	
BARREMIAN	EARLY
HAUTERIVIAN	
VALANGINIAN	
BERRIASIAN	

100–93
mya

C

CARNIVOROUS

6000kg
(5.9 tons)

NORTH
AFRICA

222

(CAR-car-oh-DONT-oh-SORE-us)

CARCHARODONTOSAURUS SAHARICUS

Carcharodon is the scientific name for the great white shark, the most fearsome killer to roam the seas. It only takes a glance at *Carcharodontosaurus* to realise why this theropod has such a name: with 20cm-long (8in), curved, serrated teeth set into a 1.5m-long (5ft) skull, and a body perhaps up to 13m (43ft) in length, it was one of the greatest carnivores ever to walk the Earth. It may have been larger than *Tyrannosaurus*, though it was less intelligent; a fine *Carcharodontosaurus* skull found in Morocco in 1995 showed that while the two dinosaurs' skulls were of similar size, *Tyrannosaurus'* braincase was 150 per cent bigger. Still, *Carcharodontosaurus* was surely a devastating killer of huge herbivores such as *Paralititan*

that lived in the coastal mangrove swamps of what is now North Africa.

The first fossils were discovered in Egypt in the 1920s by Ernst Stromer and housed, along with *Spinosaurus*, in Munich's Bavarian State Collection of Palaeontology and Historical Geology. Their destruction in wartime forced subsequent generations of experts to rely on Stromer's surviving technical drawings and descriptions. Also lost were the details of Stromer's excavation site in Egypt. But when a team from Chicago's Field Museum found the skull in Morocco and saw it fitted Stromer's description of *Carcharodontosaurus*, this cast new light on the creature. Now that it is confirmed that Stromer's excavations took place at a site called Bahariya Oasis, palaeontologists are hopeful of uncovering more of these giants in future.

**13m
(43ft)**

(EK-wee-JOO-bus)
EQUIJUBUS NORMANI

CRETACEOUS

MAASTRICHTIAN	
CAMPANIAN	
SANTONIAN	
CONIACIAN	LATE
TURONIAN	
CENOMANIAN	
ALBIAN	
APTIAN	
BARREMIAN	
HAUTERIVIAN	EARLY
VALANGINIAN	
BERRIASIAN	

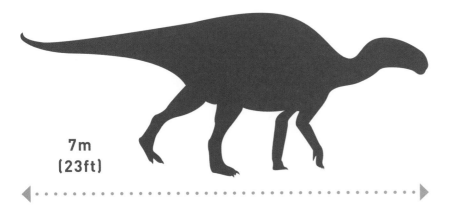

7m
(23ft)

112–98
mya

During the Early Cretaceous, *Iguanodon*-like dinosaurs began to evolve into hadrosaurs, which endured until the dinosaurs' extinction. They flourished in the Late Cretaceous, becoming the commonest herbivores in the northern hemisphere and a favourite meal for the tyrannosaurs.

Early in that transition came *Equijubus*, the most basal hadrosaur known. It had the iguanodontian's long jaws, but they were full of the small, diamond-shaped grinding teeth characteristic of the later herbivores. It also retained the ability to move both its upper and lower jaw while chewing vegetation. Whereas humans can only move the lower jaw to chew, these dinosaurs had double the grinding power.

The only known fossil is a skull and vertebrae found in 2000 in China's Ma Zong mountain range. Ma Zong means 'horse's mane', which translated into Latin becomes *Equijubus*. Like other primitive duckbills it probably retained the iguanodontian's multifunctional hands. Its spiked thumb served for defence or for feeding, an opposable 'little finger' could clutch vegetation into its palm, and three 'middle fingers' amounted to a hoof for walking on.

AS LONG
AS A
TRACTOR

HERBIVOROUS,
MIDDLE AND
LOW BROWSER

2500kg
(2.4 tons)

CHINA

223

MAASTRICHTIAN
CAMPANIAN
SANTONIAN
CONIACIAN
TURONIAN
CENOMANIAN
ALBIAN
APTIAN
BARREMIAN
HAUTERIVIAN
VALANGINIAN
BERRIASIAN

LATE

EARLY

103mya

HERBIVOROUS

27kg
(60lb)

**HWASEONG
CITY,
SOUTH KOREA**

224

(ko-REE-ah-SERR-a-tops)

KOREACERATOPS
HWASEONGENSIS

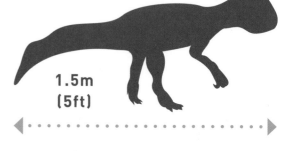

Its paddle-like tail suggests that this small ceratopsian had an aquatic lifestyle. It was not the first of those dinosaurs to prompt this speculation; *Protoceratops* has also been viewed as a possible swimmer, though other experts think that its high, flattened tail was actually used as a way of shedding excess heat or displaying to mates, so the same may apply to *Koreaceratops*. The scientists who described it in 2011 said that a more informed understanding of its behaviour would only arise if the rocks in which it was found yield information about whether it lived in an arid or coastal habitat.

Koreaceratops was the first ceratopsian found in South Korea. Its discovery filled a 20m-year gap in their presence in the fossil record, between the most primitive examples elsewhere in Asia and the famous huge varieties of Late Cretaceous North America such as *Triceratops*, which lived 35m years later. It was far smaller than most of the better known ceratopsians, around the size of a Labrador dog, and unlike them was mainly bipedal, using its hands to grasp vegetation. Its skull was not preserved but it would almost certainly have had a parrot-like face and beaked jaws.

**1.5m
(5ft)**

(DEE-a-man-TEEN-a-sore-us)

DIAMANTINASAURUS MATILDAE

................▶

**16m
(52ft)**

Two medium-sized sauropods and a deadly hunter (see overleaf) helped revolutionise Australia's reputation as a source of dinosaur remains.

For a long time the country was notoriously lacking in good fossils – most were crumbling fragmentary specimens, with a very few exceptions such as *Muttaburrasaurus* (page 221) and *Minmi* (page 194). Then in 2009 a trio of saurischians were announced together in a single paper by Australian palaeontologist Scott Hocknull and colleagues, all found in Queensland's Winton Formation rocks, which date to the late Albian stage according to an analysis of their fossil pollen

grains. Back then the excavation site was a billabong, the Australian term for a seasonal lake; today it stands by the Diamantina River and is considered a likely source of many more quality fossils.

Diamantinasaurus was a titanosaur and was unusual in bearing a prominent claw on its thumb. It probably belonged to the Saltasauridae sub-group, which had bony plates embedded over their backs as a form of armour. The specimen was nicknamed Matilda owing to the fact that the poet Andrew 'Banjo' Patterson wrote the Aussie anthem *Waltzing Matilda* in the nearby town of Winton.

AND ·········▶

CRETACEOUS

MAASTRICHTIAN	
CAMPANIAN	
SANTONIAN	LATE
CONIACIAN	
TURONIAN	
CENOMANIAN	
ALBIAN	
APTIAN	
BARREMIAN	EARLY
HAUTERIVIAN	
VALANGINIAN	
BERRIASIAN	

100mya

HERBIVOROUS

10,000kg
(9.8 tons)

**QUEENSLAND,
AUSTRALIA**

225

CRETACEOUS

MAASTRICHTIAN	
CAMPANIAN	
SANTONIAN	LATE
CONIACIAN	
TURONIAN	
CENOMANIAN	
ALBIAN	
APTIAN	
BARREMIAN	EARLY
HAUTERIVIAN	
VALANGINIAN	
BERRIASIAN	

100mya

C

CARNIVOROUS

500kg
(1100lb)

QUEENSLAND,
AUSTRALIA

226

(OS-tray-LO-ven-ayt-or)

AUSTRALOVENATOR WINTONENSIS

AND

**6m
(20ft)**

The song describes a swagman who drowns in a billabong. Around 100mya this seems to have been the fate of *Australovenator*, a speedy carnivore whose fossil was nicknamed Banjo. *Australovenator* was probably scavenging the already-dead *Diamantinasaurus'* carcass at the waterside when both grew engulfed in mud and eventually sank to the bottom of the billabong, only to emerge in the 21st century from the clay sediments.

Australovenator was an exciting discovery. It was the most complete theropod skeleton yet found in Australia, a basal carcharodontosaur related to immense carnivores such as *Giganotosaurus* that terrorised Late Cretaceous South America. *Australovenator* was smaller than them but still the size of a horse, and a lithe and energetic hunter. According to Cambridge palaeontologist Roger Benson and colleagues it was a member of Neovenatoridae (see page 173).

(win-TOE-no-tie-tan)

WINTONOTITAN WATTSI

Wintonotitan's discovery has a different story, having been collected in 1974 by Dr Keith Watts and originally assigned to *Austrosaurus*. In reassessing it Dr Hocknull likened its build to a hippopotamus' whereas *Diamantinasaurus*' was closer to a giraffe's. Like the other hefty sauropod it probably had an armoured back. Both were small by sauropod standards – some species were ten times their weight – and both roamed the Gondwanan plains feasting on vegetation from gingko leaves to cycads and ferns, all of which are fossilised in the Winton rock formation. The research team nicknamed it Clancy, which references another Patterson song.

Put together, Banjo, Matilda and Clancy suddenly added colour and diversity to a vast southern swathe of the ancient world that is at last yielding its secrets. It's known that far bigger sauropods roamed Australia, though. There are hundreds of fossil footprints preserved along the Broome coastline in Western Australia, some of them a staggering 1.5m (5ft) in diameter. Will the dinosaurs that left them emerge from the rock strata soon?

MAASTRICHTIAN
CAMPANIAN
SANTONIAN
CONIACIAN
TURONIAN
CENOMANIAN
ALBIAN
APTIAN
BARREMIAN
HAUTERIVIAN
VALANGINIAN
BERRIASIAN
LATE
EARLY

100mya

HERBIVOROUS

10,000kg
(9.8 tons)

QUEENSLAND, AUSTRALIA

15m
(49ft)

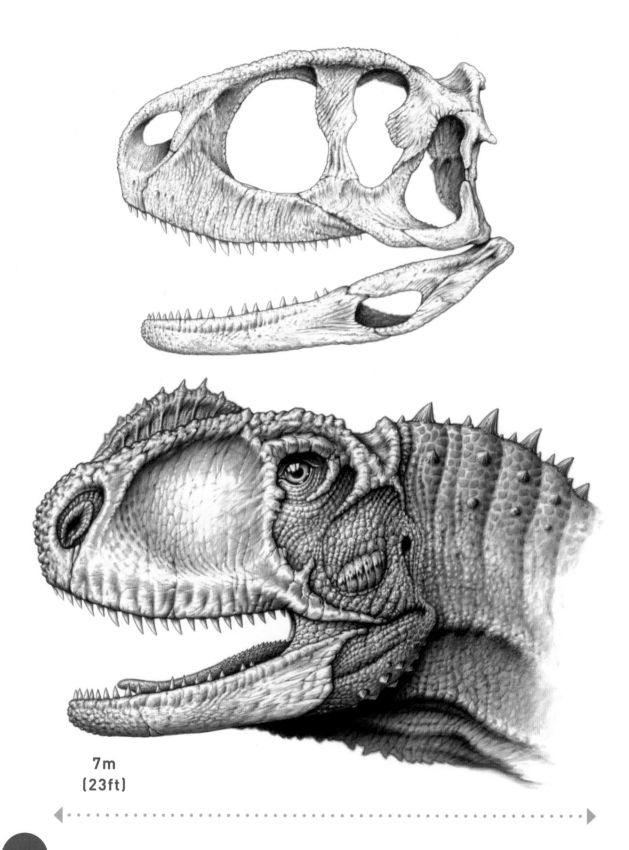

7m
(23ft)

(ROO-gops)
RUGOPS PRIMUS ·········· ▶

CRETACEOUS

MAASTRICHTIAN	
CAMPANIAN	
SANTONIAN	LATE
CONIACIAN	
TURONIAN	
CENOMANIAN	
ALBIAN	
APTIAN	
BARREMIAN	EARLY
HAUTERIVIAN	
VALANGINIAN	
BERRIASIAN	

With only a partial skull discovered, this hefty abelisaurid's appearance is hard to gauge, but the fossil provides interesting pointers: the face has two rows of seven holes that may have contained spikes, and the fact that it is relatively delicate and riddled with blood vessels hints that *Rugops* (meaning 'wrinkle face') was a scavenger rather than a killer. Dinosaurs that bit and killed other animals tended to have heavier skulls.

Like most other abelisaurids, it probably had tiny arms of uncertain purpose – one possibility is that they had some sexual display role. However, the most important fact about *Rugops'* discovery was that it clarified understanding of how and when Gondwana split and drifted into the arrangement that we know today. This southern super-continent combined what we now call Africa, Madagascar, South America and Asia. Before *Rugops'* discovery in Niger in 2000, abelisaurids were unknown in Africa, suggesting that Africa had already split from the rest of Gondwana by the time of their emergence around 110mya. The discovery of the *Rugops'* skull dating from around 95mya implies that Africa was still attached to the rest of Gondwana at that stage.

95mya

CARNIVOROUS

750kg
(1650lb)

ABANGHARIT,
NIGER

229

CRETACEOUS

MAASTRICHTIAN	
CAMPANIAN	
SANTONIAN	LATE
CONIACIAN	
TURONIAN	
CENOMANIAN	
ALBIAN	
APTIAN	
BARREMIAN	EARLY
HAUTERIVIAN	
VALANGINIAN	
BERRIASIAN	

99–83
mya

C

CARNIVOROUS

250kg
(550lb)

**DORNOGOV,
MONGOLIA**

230

(a-KILL-oh-BAHT-er)

ACHILLOBATOR GIGANTICUS

This feathered predator had a lethal, sickle-like claw on each foot, which would have required a powerful Achilles tendon to act as an effective weapon in killing other dinosaurs. *Achillobator* was a 2m (6ft 6in) member of the Dromaeosauridae, aggressive birdlike carnivores prevalent worldwide from the Middle Jurassic to the Late Cretaceous. It was one of the later, more advanced examples – current thinking places it as a close relation to *Utahraptor* (page 178), occupying a different branch from *Velociraptor* (page 282) and *Deinonychus* (page 213). There is no evidence from the fossil remains, which were discovered in Mongolia in 1989, but it is assumed that *Achillobator* bore a covering of dino-fuzz.

**HUGE
FEATHERED
PREDATOR**

6m
(19ft)

(OR-ick-toe-DRO-me-us)

ORYCTODROMEUS CUBICULARIS

CRETACEOUS

MAASTRICHTIAN	
CAMPANIAN	
SANTONIAN	LATE
CONIACIAN	
TURONIAN	
CENOMANIAN	
ALBIAN	
APTIAN	
BARREMIAN	EARLY
HAUTERIVIAN	
VALANGINIAN	
BERRIASIAN	

2.1m
(7ft)

95mya

H

HERBIVOROUS

20kg
(44lb)

This small herbivore is one of just a few burrowing dinosaurs. Three fossils were found in a 2m-long, 70cm-wide (6ft 6in by 2ft 4in) pocket of sandstone that lay within a formation of mudstone in western Montana. The dinosaurs died within a burrow they had dug into the muddy earth, and then wind blew the burrow full of sand, which turned to stone and enclosed them for 95m years. *Oryctodromeus* was a fast-running biped that is thought to have had an unusually flexible tail, allowing it to turn around in confined spaces.

MONTANA,
USA

231

CRETACEOUS

MAASTRICHTIAN	
CAMPANIAN	
SANTONIAN	LATE
CONIACIAN	
TURONIAN	
CENOMANIAN	
ALBIAN	
APTIAN	
BARREMIAN	EARLY
HAUTERIVIAN	
VALANGINIAN	
BERRIASIAN	

99–84 mya

H

HERBIVOROUS

uncertain

OMNOGOV,
MONGOLIA

(GRA-sill-ee-SERR-a-tops)

GRACILICERATOPS MONGOLIENSIS

**90cm (3ft) as a juvenile;
probably 2m (6ft 6in) as an adult**

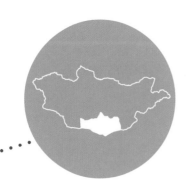

This lean, limber biped was a little relative of the mighty rhinoceros-like four-legged ceratopsians of the Late Cretaceous. The partial remains – a skull and some disarticulated bones – found in Mongolia included a hind limb longer than the front, suggesting it was an upright scurrying runner. In common with other ceratopsians it had a parrot-like beak for cutting through vegetation (probably ferns, cycads and conifers) and a thick protective plate at the back of the skull, though nothing as spectacular as the frilled shields developed by *Triceratops* or *Centrosaurus*. After the fossil's discovery in 1975 *Graciliceratops* was initially considered an adult specimen of the already-known *Microceratops*, but in 2000 the American palaeontologist Paul Sereno identified it as a juvenile of a new genus.

(MAP-you-SORE-us)
MAPUSAURUS ROSEAE

MAASTRICHTIAN
CAMPANIAN
SANTONIAN
CONIACIAN
TURONIAN
CENOMANIAN
ALBIAN
APTIAN
BARREMIAN
HAUTERIVIAN
VALANGINIAN
BERRIASIAN

LATE

EARLY

Imagine a huge carnivore possibly the size of *Tyrannosaurus* pounding after its prey... and then picture a pack of half-a-dozen all hunting together. Quite apart from growing to an immense size, what is most exciting about *Mapusaurus* is the hint that it pulled down huge sauropods by working as a team.

In 1997 palaeontologists working in western Argentina found the bones of at least seven of this previously unknown carcharodontosaur, all at different stages of maturity. If they were social animals then this could explain how they managed to eat enough meat to survive: for while *Mapusaurus* was huge for a theropod, it was dwarfed by the sauropod that formed its main prey. *Argentinosaurus* (page 236) was the bulkiest of all known dinosaurs at 30–35m (100ft) long and perhaps 75 tonnes (74 tons) in weight. A single *Mapusaurus* would have had little effect – in fact they may have bitten chunks out of *Argentinosaurus* for a snack, leaving the victim to lumber on hurt but ultimately unaffected.

However, a group of *Mapusaurus* working as a team could easily have brought down a juvenile *Argentinosaurus* and quite possibly an adult too. Describing the find in 2006, palaeontologists Rodolfo Coria and Philip Currie noted that, like its close relatives *Giganotosaurus* and *Carcharodontosaurus*, it had bladed teeth made for slicing through meat, whereas the conical teeth of the later *Tyrannosaurus* were for crushing bone. *Argentinosaurus'* bones were simply too big to crunch through – but a pack simultaneously biting out lumps of flesh all over its body would probably bring it to ground, whereupon the fatal bites could be inflicted.

Similar apparent predator–prey pairings were widespread: wherever there was a huge carcharodontosaur, palaeontologists have also found evidence of a colossal herbivore. So just like South America's *Mapusaurus* and *Argentinosaurus*, in Africa *Carcharodontosaurus* perhaps preyed on *Paralititan*, and in North America, *Acrocanthosaurus* seemingly hunted *Sauroposeidon*. When the herbivores died out, the carnivores followed. Living around 30m years before the great extinction, the enormous pairing of *Mapusaurus* and *Argentinosaurus* represents the apex of the dinosaurs' development in terms of size, if not sophistication. From this point everything began gradually to diminish in scale.

99–93 mya

C
CARNIVOROUS

5000kg (4.9 tons)

ARGENTINA

11.5m (38ft)

(illustrated overleaf)

233

MAPUSAURUS
ROSEAE

A pack of *Mapusaurus* attack a pair of *Argentinosaurus* – these dinosaurs are an example of a predator–prey relationship between carcharodontosaurs and sauropods.

CRETACEOUS

MAASTRICHTIAN	
CAMPANIAN	
SANTONIAN	LATE
CONIACIAN	
TURONIAN	
CENOMANIAN	
ALBIAN	
APTIAN	
BARREMIAN	EARLY
HAUTERIVIAN	
VALANGINIAN	
BERRIASIAN	

96–94
mya

HERBIVOROUS

75,000kg
(73.8 tons)

NEUQUEN
PROVINCE,
ARGENTINA

236

(ar-jen-TEEN-oh-SORE-us)

ARGENTINOSAURUS HUINCULENSIS

As tall as a six-storey building, as long as three buses, as heavy as half a dozen elephants... this gargantuan sauropod is the biggest land animal known for sure to have existed.

Argentinosaurus was the perfect herbivorous eating machine: everything about it was intended to help it eat as much as possible, as quickly as possible, extracting the maximum energy possible from what it ate while expending very little in doing so. Its long neck let it reach high, low and wide to crop leaves from trees and bushes while keeping its heavy legs planted on the ground. It didn't waste time or effort chewing, it just swallowed the leaves whole and let the bacteria in its gut break the fibres down. Not needing heavy teeth for chewing meant its head was light and easy to support. And so it

evolved ever bigger, doing so in a symbiotic relationship with *Mapusaurus* (page 233).

Its discovery began in 1988 when a Patagonian sheep farmer named Guillermo Heredia found what looked like a petrified tree trunk on his land. But the more he peered at it, the more it intrigued him... and when he called in a team of palaeontologists from the Carmen Funes Municipal Museum they concluded that it was a 1.5m (5ft) shinbone. Further excavations took time – it took five men to haul out a mighty lump of rock that turned out to contain a single vertebra – but eventually a few more vertebrae, fractured ribs and the sacrum were extracted from the Huincul Formation stone, enough to gauge the animal's size. Jose Bonaparte and Ricardo Coria named *Argentinosaurus* in 1993 and a general public raised on the idea that *Brachiosaurus* and *Apatosaurus* were the biggest dinosaurs had to think again.

30m
(98ft)

MYSTERIOUS GIANTS

• • • • Edward Drinker Cope,
1840–1897

Argentinosaurus was huge – but there are tantalising suggestions of animals that left it looking distinctly diminutive. *Amphicoelias fragillimus* is one of the mysteries of dinosaur research: the incredibly large diplodocid whose single known bone was discovered, described and then lost. In 1877 one of Edward Drinker Cope's fossil collectors found a partial vertebra in Morrison Formation rocks in Colorado. It was 2.7m (89ft) tall – that is, a single piece of backbone the height of an elephant. This suggests an animal that might have been 60m (200ft) long and weighed 110 tonnes (108 tons).

Cope drew and described the find… and then it vanished from trace. No one knows what happened. The best guess is that being a fragile fossil recovered from crumbling, 150m-year-old Jurassic mudstone, it simply disintegrated while in storage and Cope disposed of it. Now the only remaining evidence is his single drawing of the bone, and that is not sufficient evidence for *Amphicoelias* to be confirmed as the biggest dinosaur known.

It had a heavier though less lengthy rival in *Bruhathkayosaurus matleyi*, which was announced more recently but is still considered dubious. Discovered in

southern India and described in 1989, the remains seemingly comprised hip bones, front and rear leg bones and a section of tail. According to the researchers the shin bone measured 2m (6ft 6in) and the humerus 2.3m (7ft 6in); compare that with *Argentinosaurus'* 1.5m (5ft) and 1.8m (6ft) respectively. Working from these figures, it's proposed that *Bruhathkayosaurus* was up to 34m (112ft) long and weighed up to 139 tonnes (137 tons).

Although it's awe-inspiring to picture these 'mega-sauropods' as they're termed, there's insufficient scientific substance for palaeontologists to get too excited. They are loath to confirm *Bruhathkayosaurus* as a genus because of the lack of analysis, description and scientific drawing offered by the scientists who announced it. There's even a possibility that these remains aren't bones at all, but chunks of petrified wood. Most frustratingly, they were never fully extracted from their surrounding rock and are believed to have since been severely eroded by the weather.

Put together though, *Amphicoelias* and *Bruhathkayosaurus* do raise suspicions that the mightiest sauropods whose existence is confirmed were only middle-ranking members of a world of giants.

237

CRETACEOUS

MAASTRICHTIAN	
CAMPANIAN	
SANTONIAN	**LATE**
CONIACIAN	
TURONIAN	
CENOMANIAN	
ALBIAN	
APTIAN	
BARREMIAN	**EARLY**
HAUTERIVIAN	
VALANGINIAN	
BERRIASIAN	

93mya

C

CARNIVOROUS

1700kg
(1.6 tons)

**WESTERN
ARGENTINA**

238

(SCOR-pee-oh-ven-AYT-or)

SKORPIOVENATOR BUSTINGORRYI

This gnarled, ugly predator roamed the woodlands of Late Cretaceous Argentina, vying for prey with other big killers such as *Mapusaurus* and *Ilokelesia*. Its diet included the sauropod *Cathartesaura*, and not scorpions in case you wondered – the name refers to the colony of living scorpions teeming all over the palaeontologists' excavation site, which lay within a Patagonian farm owned by a Manuel Bustingorry. It was a risky dig but well worth it, as they left with the near-complete skeleton of a hitherto unknown major theropod.

The ridges between its eyes mark *Skorpiovenator* out as an abelisaurid, the group of carnivores prevalent around Late Cretaceous South America and to a lesser extent Africa. Its blunt skull led to its classification in a sub-group called Brachyrostra, or 'short snouts'. In publishing their description in 2008, the Argentinian team of experts led by Juan Canale noted that the 'skull shortening and hyperossification [thickening of bone] of the skull roof' are 'possibly related to shock-absorbing capabilities', which suggests that these aggressive, truck-sized beasts habitually charged at and headbutted each other.

**7.5m
(25ft)**

(JIG-ah-NOT-oh-SORE-us)

GIGANOTOSAURUS CAROLINII

13m
(43ft)

CRETACEOUS

MAASTRICHTIAN	
CAMPANIAN	
SANTONIAN	LATE
CONIACIAN	
TURONIAN	
CENOMANIAN	
ALBIAN	
APTIAN	
BARREMIAN	EARLY
HAUTERIVIAN	
VALANGINIAN	
BERRIASIAN	

97mya

CARNIVOROUS

8000kg
(7.8 tons)

ARGENTINA

239

With a skull the length of a man and a body as long as a bus, this is perhaps the biggest predatory carnivorous dinosaur ever known – some experts believe that it grew bigger than *Tyrannosaurus*, though others are more conservative and suspect that both genera's biggest individuals were of similar size. In any case size is one thing, intelligence quite another. Although palaeontologists don't have much time for such questions, it is tempting to wonder which dinosaur would win in a battle between *Giganotosaurus* and *Tyrannosaurus*. Even if we accept the idea that *Giganotosaurus* was longer, two tonnes heavier and had bigger jaws (which were full of teeth shaped for slicing flesh), *Tyrannosaurus* had a bite three times as powerful, broader and more varied bone-crunching teeth, and a bigger brain. The question is entirely speculative in any case, as the two dinosaurs lived 30m years apart and on separate continents. *Giganotosaurus'* discovery in Argentina showed that South America was not only the land where dinosaurian life may have begun, it was also where it reached its pinnacle in terms of size. Like its fellow massive carcharodontosaurids, *Giganotosaurus* lived alongside gigantic herbivores; its fossils were found in 1993 by amateur fossil-hunter Ruben Dario Carolini near those of the titanosaurid *Andesaurus* and rebbachisaurid *Limaysaurus*. Like its relative *Mapusaurus*, *Giganotosaurus* (the 'giant southern lizard') may have hunted these mammoth beasts.

LIVING FOSSILS

'These anomalous forms may almost be called living fossils; they have endured to the present day, from having inhabited a confined area, and from having thus been exposed to less severe competition.'

Charles Darwin, *On the Origin of Species*

Birds form a lineage that stretches all the way back to the age of the dinosaurs – but while the connection is clear to see, they have evolved a good deal since that time. Perhaps even more astonishing are the animals and plants that lived alongside the dinosaurs and still exist in the world today, having barely changed in hundreds of millions of years. Here are five of these amazing survivors.

COELACANTH
(*Latimeria chalumnae*)

Varieties of this large bony fish swam the seas long before the dinosaurs' time on Earth and they continue to do so today. When coelacanth (pronounced SEE-la-canth) fossils were found in Australia in the mid-19th century they were dated to around 360mya. Subsequent fossil finds led scientists to believe that the coelacanth had died out in the Cretaceous-Tertiary extinction around 65mya. Then in 1938, Marjorie Latimer, the curator of a small South African museum, spotted an odd-looking fin jutting out from among a pile of rays and sharks on a fishing trawler that had come into dock at East London, near Cape Town. She ignored the advice of those who told her it was a variety of cod and, after studying reference books that contained examples of uncannily similar prehistoric fish, she alerted a university professor, J L B Smith, who confirmed its importance. Soon it was proclaimed the 'zoological find of the century', no less remarkable than discovering a living dinosaur.

Coelacanths have since been found – and recently filmed – at numerous places along the Indian Ocean coastlines of eastern Africa and Indonesia (those in Indonesia being from a second species, *L. menadoensis*). They spend the daytime lurking in caves around 200m (650ft) below the surface and emerge at night to feed on other fish, using a uniquely hinged jaw that can open extremely wide. Coelacanths grow up to 1.8m (6ft) long, are covered in tough scales that serve as armour and, most immediately noticeable, have eight fins including a parallel pair on the back. This gives them unusual manoeuvrability – they have been filmed swimming upside down and even in a headstand position.

HORSESHOE CRAB
(*Limulus polyphemus*)

Primitive creatures such as trilobites and horseshoe crabs were among the earliest life-forms to scuttle along the beds of the shallow seas of the Palaeozoic era around 450mya. The trilobites died out in the Permian extinction event 250mya and are only known from ancient fossils, but the horseshoe crab survived and is still with us, having hardly altered since then.

Found today in locations including the USA and Japan, it is not actually a crab but a form of arthropod more closely related to spiders and scorpions. Beneath its rounded shell it looks similar to a large scorpion with an extra pair of legs. It has nine eyes, including two compound eyes similar to those of flies, and several light-sensors along its tail. Stranger still, its blood is a blue-green colour. This is because it contains a copper-based pigment called haemocyanin, rather than the haemoglobin that colours human blood red.

The horseshoe crab is considered at risk in some locations because its habitat is threatened and because of over-harvesting – it is often used as fishing bait. It is a vital source of food for endangered sea turtles, and is also harvested for medical reasons, which provides another strong reason to safeguard its future. Its unusual blood contains a protein called Limulus Amebocyte Lysate (LAL), which pharmaceutical companies use to test their products for endotoxins, bacterial substances that prove fatal to humans. This is considered one of the most important medical products obtained from a sea creature.

241

THE GINKGO TREE
(*Ginkgo biloba*)

Voracious herbivores such as *Iguanodon* gorged on the ginkgo, among many other forms of vegetation. Today the tree's distinctive leaves remain identical to those preserved in fossil impressions from the Early Jurassic, and they are recognisably similar to examples from all the way back to the Permian age, 270mya. Also known as the maidenhair tree on account of its oddly shaped leaves, the ginkgo is a unique survivor from a time before flowers even existed. It grows up to 30m (100ft) tall and is dioecious, meaning that individual trees are either male or female. The ginkgo grows in the wild in China but is a popular ornamental tree in parks and gardens worldwide. Male specimens are usually preferred because the females produce fleshy yellow seeds that produce an unpleasant smell similar to rancid butter. Ginkgo trees can easily reach 1000 years old and the oldest known example in China is said to be aged 3500. It's also the source of a popular herbal remedy that some people think aids memory and mental sharpness.

TUATARA
(*Sphenodon punctatus*)

Found only in New Zealand, this reptile might look like a lizard but in fact is the only survivor of the group Sphenodontida, which was widespread around 200mya. The tuatara has a 'third eye', a light sensitive patch on the top of its head, the purpose of which may be to absorb Vitamin D from sunlight or to regulate its body cycles according to the time of day.

Some scientists dispute whether the tuatara merits the term 'living fossil' because it has evolved since the dinosaurs' time. But there is no doubt that it is the last survivor of a group stretching back to the late Permian age, thus actually predating the dinosaurs. This explains some remarkable aspects of its anatomy. It has no earholes, although it can hear owing to the presence of two primitive receptors hidden within its head, and it has unusual bony ridges on its skull and a unique arrangement of teeth: two rows on its upper jaw that fit around a single row on the lower jaw. It is a nocturnal hunter of insects and spiders, grows to around 80cm (2ft 7in) in length and can live to be more than 100 years old – and keep breeding as long as it lives! In 2009 a tuatara kept at a New Zealand zoo became a father at the age of 111.

GOBLIN SHARK
(*Mitsukirina owstoni*)

Some very strange looking fish lurk in the oceans' lightless depths and they don't come much uglier than the goblin shark, which hunts for squid, crabs and fish around 250m (820ft)beneath the surface – though the deepest ever found was an incredible 1300m (4265ft) underwater. From where you are sitting imagine somewhere almost a mile away: and then imagine that distance downwards through murky, cold seawater. The fossil record shows that its appearance has changed little since the Late Cretaceous. It grows to around 3m (10ft) long and has a pinkish skin, but its most notable feature is the long protuberance from its head, which emits an electric field enabling it to track its prey in the darkness. The shark then suddenly extends its jaws and sucks its prey into its mouth. Goblin sharks are occasionally caught by fishing vessels but they rarely survive long in captivity, as they suffer from decompression after being removed from the extreme water-pressure of the deep seas.

CRETACEOUS

MAASTRICHTIAN	
CAMPANIAN	
SANTONIAN	LATE
CONIACIAN	
TURONIAN	
CENOMANIAN	
ALBIAN	
APTIAN	
BARREMIAN	
HAUTERIVIAN	EARLY
VALANGINIAN	
BERRIASIAN	

90mya

H

HERBIVOROUS

50,000kg
(49 tons)

NEUQUEN
PROVINCE,
ARGENTINA

244

(FOOT-ah-LONK-oh-SORE-us)

FUTALOGNKOSAURUS DUKEI

This gigantic titanosaur's remains were hailed as the most complete giant sauropod skeleton ever found at the announcement of its discovery in 2007. Whereas fellow titanosaur *Argentinosaurus* is known from only 3 per cent of a skeleton, the *Futalognkosaurus* specimen comprised around 70 per cent. Its gargantuan bones were found at Futalognko, a densely stocked fossil site in Argentina, and included an entire neck the length of a bus and a metre deep. As in *Isisaurus* (page 304), this strange neck may have functioned as a form of advertising hoarding decorated with markings to attract a mate.

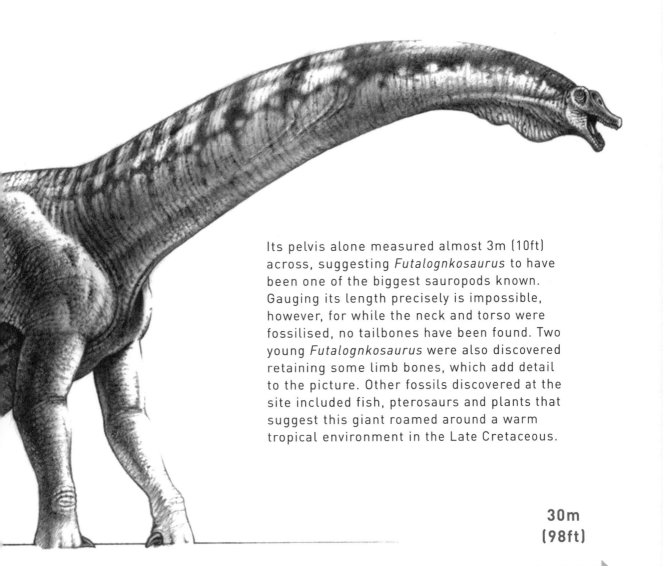

Its pelvis alone measured almost 3m (10ft) across, suggesting *Futalognkosaurus* to have been one of the biggest sauropods known. Gauging its length precisely is impossible, however, for while the neck and torso were fossilised, no tailbones have been found. Two young *Futalognkosaurus* were also discovered retaining some limb bones, which add detail to the picture. Other fossils discovered at the site included fish, pterosaurs and plants that suggest this giant roamed around a warm tropical environment in the Late Cretaceous.

30m (98ft)

MAASTRICHTIAN	
CAMPANIAN	
SANTONIAN	LATE
CONIACIAN	
TURONIAN	
CENOMANIAN	
ALBIAN	
APTIAN	
BARREMIAN	
HAUTERIVIAN	EARLY
VALANGINIAN	
BERRIASIAN	

94–89
mya

HERBIVOROUS

150kg
(330lb)

NEW MEXICO,
USA

246

(zoo-nee-SERR-a-tops)

ZUNICERATOPS CHRISTOPHERI

3m
(10ft)

Some palaeontologists spend their whole careers digging and scraping at ancient rock formations around the planet without managing to uncover a significant new genus of dinosaur. Others pitch up in New Mexico at the age of eight, start digging and find a *Zuniceratops*. It was Christopher James Wolfe, the son of palaeontologist Douglas Wolfe, who managed that feat in 1996. The skull and few bones he found were enough to fill in a gap in our knowledge of the ceratopsians' evolution: this metre-tall (3ft) dinosaur was the earliest one known to have horns sprouting from above its eyes, predating the better known *Triceratops* by 10m years. It had only two of them, whereas many later examples had an extravagant assortment of horns and spikes. Its name means 'Zuni-horned-face', Zuni referring to the local Native American people.

(MEG-ah-rap-tor)
MEGARAPTOR NAMUNHUAIQUII · · · · · · · · ▷

A 'raptor' with a 30cm-long (1ft) killing claw? That was the original view prompted by the retrieval of a lethal hook from an Argentine rock formation, but the later discovery of a whole forelimb showed that it belonged on the hand of this monstrous allosauroid. The result was a prospect no less terrifying: a killing machine wielding claws even more devastatingly scythe-like than the spinosaurids.

Of as much interest to palaeontologists is what this South American dinosaur tells them about the break-up of Gondwana. The discovery of two very similar theropods to this in Australia – *Australovenator* and another that's as yet unnamed – suggest that that country and South America remained connected later than previously believed.

CRETACEOUS

MAASTRICHTIAN	
CAMPANIAN	
SANTONIAN	LATE
CONIACIAN	
TURONIAN	
CENOMANIAN	
ALBIAN	
APTIAN	
BARREMIAN	EARLY
HAUTERIVIAN	
VALANGINIAN	
BERRIASIAN	

90mya

C

CARNIVOROUS

1000kg
(0.9 tons)

8m
(26ft)

RIO NEUQUEN
FORMATION,
PATAGONIA,
ARGENTINA

(NEW-kwen-rap-tor)

NEUQUENRAPTOR ARGENTINUS

2.5m (8ft)

CRETACEOUS

MAASTRICHTIAN	
CAMPANIAN	
SANTONIAN	LATE
CONIACIAN	
TURONIAN	
CENOMANIAN	
ALBIAN	
APTIAN	
BARREMIAN	EARLY
HAUTERIVIAN	
VALANGINIAN	
BERRIASIAN	

93–85 mya

C

CARNIVOROUS

30kg (66lb)

NEUQUEN PROVINCE, ARGENTINA

Deinonychosaurs were only known from the former Laurasia in the northern hemisphere until *Unenlagia* and *Neuquenraptor*'s discovery in the mid-1990s showed that Gondwana was also home to these bird-like predators. *Neuquenraptor*'s few fragmented remains included a footbone with the tell-tale killing claw that showed this to be a close relative of *Unenlagia*, and perhaps even the same animal. Given that Laurasia and Gondwana had been split for millions of years by the Late Cretaceous, their emergence in western Argentina showed that deinonychosaurs must have had an ancient heritage predating the landmasses' division.

AND

249

CRETACEOUS

MAASTRICHTIAN	
CAMPANIAN	
SANTONIAN	LATE
CONIACIAN	
TURONIAN	
CENOMANIAN	
ALBIAN	
APTIAN	
BARREMIAN	EARLY
HAUTERIVIAN	
VALANGINIAN	
BERRIASIAN	

94–86
mya

C

CARNIVOROUS

75kg
(165lb)

ARGENTINA

(OON-en-LAHG-ee-ah)

UNENLAGIA COMAHUENSIS

Other species: *U. patnemili*

Unenlagia was initially taken for a form of ancient bird, with its wing-like arms and shoulder joints that would have permitted a flapping motion. It was granted a name meaning 'half-bird' but whether it was any sort of bird is debatable. It was a flightless predator that appears to have evolved from flying ancestors: studies suggest that the flightless dromaeosaurids of the Late Cretaceous descended from Jurassic avian dinosaurs such as *Archaeopteryx*, with which they share a very similar pelvic structure. As they evolved to grow larger and heavier – *Unenlagia* was at least 3m (10ft) long – they became too big to fly, as is the case with ostriches.

So this would suggest that animals within this lineage evolved to fly twice: first back in Jurassic times, then in the form of the modern birds into which dromaeosaurids' relatives eventually developed. *Unenlagia*'s name derives from the isolated Mapudungun language spoken in part of Argentina, where a single specimen of 20 bones was removed from sedimentary river rocks.

3.5m (11ft)

(ant-ARC-toe-SORE-us)

ANTARCTOSAURUS WICHMANNIANUS

18m
(60ft)

CRETACEOUS

MAASTRICHTIAN	
CAMPANIAN	
SANTONIAN	LATE
CONIACIAN	
TURONIAN	
CENOMANIAN	
ALBIAN	
APTIAN	
BARREMIAN	EARLY
HAUTERIVIAN	
VALANGINIAN	
BERRIASIAN	

83–80 mya

This wide-mouthed sauropod was probably 18m long and three times an average human's height at the shoulder, possibly covered in armour and could have weighed 33.5 tonnes (3 tons). *Antarctosaurus* may have had another species, *A. giganteus*, that rivalled *Argentinosaurus* in stature. Nothing more definite can be said until a better fossil is discovered; the present notion of its appearance is extrapolated from a few disconnected bones found in the Castillo Formation in Argentina. This hints at another layer of confusion where *Antarctosaurus* is concerned: it didn't live in the Antarctic but in the area that is now Argentina, Chile and Uruguay. The name, coined in 1929 by the prolific palaeontologist Friedrich von Huene, simply means 'non-northern lizard' – which we'll grant is accurate, given that it lived in the southern hemisphere, but it couldn't be called a vintage moment in the history of dinosaur nomenclature. The first known dinosaur that did live in what's now the Antarctic was only discovered in 1986 and its name, less misleadingly, is *Antarctopelta*.

H

HERBIVOROUS

33,500kg (33 tons)

ARGENTINA

251

MAASTRICHTIAN	
CAMPANIAN	
SANTONIAN	LATE
CONIACIAN	
TURONIAN	
CENOMANIAN	
ALBIAN	
APTIAN	
BARREMIAN	EARLY
HAUTERIVIAN	
VALANGINIAN	
BERRIASIAN	

75mya

OMNIVOROUS

22kg
(50lb)

MONGOLIA

252

(OH-vee-rap-tor)

OVIRAPTOR PHILOCERATOPS

This most unfair of dinosaur names means 'egg thief, lover of ceratopsians', and it was conferred because American researchers on the 1923 Mongolian expedition that unearthed *Velociraptor* discovered this theropod apparently caught in the act of raiding a *Protoceratops'* nest. What's more, its toothless, sharp-rimmed beak seemed designed for snapping through eggshells. But even as he announced its name, expedition leader Henry Fairfield Osborn conceded that it might prove misleading – and he was right, for this feathered bird-like dinosaur was actually sitting by its own eggs, not stealing another's. This was proved by the discovery of the very similar *Citipati* brooding on a nest of eggs identical to those found beside *Oviraptor*. The Mongolian palaeontologist Rinchen Barsbold suggested in the 1970s that *Oviraptor*'s distinctive jaws were instead designed for crunching through molluscs' shells, as clams are found fossilised within the same Djadochta rock formation.

However, the single confirmed *Oviraptor* fossil does contain the bones of a small lizard in its stomach, suggesting it ate whatever it could. That being the case, it's even possible that among many other things it did eat other animals' eggs – but there is no reason to think that they formed the basis of its diet.

AS LONG AS AN AVERAGE BROOM

1.5m
(5ft)

(dee-AB-loh-SERR-a-tops)

DIABLOCERATOPS EATONI

CRETACEOUS

MAASTRICHTIAN	
CAMPANIAN	LATE
SANTONIAN	
CONIACIAN	
TURONIAN	
CENOMANIAN	
ALBIAN	
APTIAN	
BARREMIAN	EARLY
HAUTERIVIAN	
VALANGINIAN	
BERRIASIAN	

5.5m
(18ft)

78mya

H

HERBIVOROUS

2000kg
(1.9 tons)

Magnificent horns sprouted from the brows and bony frill of this ceratopsian, which despite its ornate decoration is one of the most primitive yet discovered. Donald DeBlieux found its slender, metre-long (3ft) skull in Utah's Grand Staircase Escalante National Monument in 2002. It sat there for years in a heavy block of sandstone until he secured free helicopter transportation to airlift it out of the badlands and into a workshop. After 800 hours' work sawing away the stone he had a beautifully preserved skull to work from, and in 2010 he named and described it with fellow palaeontologist James Kirkland.

UTAH, USA

253

MAASTRICHTIAN	
CAMPANIAN	LATE
SANTONIAN	
CONIACIAN	
TURONIAN	
CENOMANIAN	
ALBIAN	
APTIAN	
BARREMIAN	EARLY
HAUTERIVIAN	
VALANGINIAN	
BERRIASIAN	

78mya

HERBIVOROUS

(ZEE-no-SERR-a-tops)

XENOCERATOPS FOREMOSTENSIS

6m
(20ft)

2000kg
(1.9 tons)

FOREMOST,
ALBERTA,
CANADA

Discovering dinosaurs isn't always a case of chipping away at a rockface out in the badlands – sometimes it just involves opening long-forgotten boxes in a museum's archives and seeing what lies within. That is what Michael Ryan and David Evans did and in 2012 they described the collected fragments from three individuals as a new genus of centrosaurine ceratopsid. Dr Ryan, of the Cleveland Museum of Natural History, and Dr Evans, of the Royal Ontario Museum, discovered the bones at the Canadian Museum of Nature in Ottawa as part of their Southern Alberta Dinosaur Project, which aims to fill in gaps in Canada's fossil record. Though the first part of its name means 'alien' (as in xenophobia, or the fear of foreigners), this did not refer to

Xenoceratops' strange appearance. With a parrot-like beak, two long brow horns above its eyes, and a large frill bearing two great spikes, it looks no more or less strange than any other of the many-horned herbivores that populated western North America in the Cretaceous. Instead it refers to the scarcity of horned dinosaur fossils in the Canadian rocks where it was fossilised. American palaeontologist Wann Langston Jr retrieved the skull fragments in 1958 from the Foremost Formation, which is in Alberta. Other nearby rocks such as the Dinosaur Park Formation are famed for their rich crop of fossils but the Foremost Formation rocks have so far yielded very little. These strata are around 78m years old, making *Xenoceratops* the earliest ceratopsian known from Canada.

(ZOO-cheng-tie-RAN-us)
ZHUCHENGTYRANNUS MAGNUS

CRETACEOUS

MAASTRICHTIAN	
CAMPANIAN	
SANTONIAN	LATE
CONIACIAN	
TURONIAN	
CENOMANIAN	
ALBIAN	
APTIAN	
BARREMIAN	EARLY
HAUTERIVIAN	
VALANGINIAN	
BERRIASIAN	

There aren't many places more exciting for dinosaur researchers than China right now. Every year seems to bring revelatory finds, such as *Zhuchengtyrannus*, a major predator whose discovery was announced in 2011.

The name refers to a particular hotspot for fossil-hunters: the city of Zucheng, known locally as 'dinosaur city'. It has been deemed an important site since 1960 but gained new prominence in December 2008 when local palaeontologists announced that they had unearthed the remains of 7600 creatures within the previous nine months and claimed Zucheng as the world's biggest fossil field. The bones date from the very end of the dinosaurs' reign, just shy of the K-T Boundary that marks the non-avians' extinction, and it is hoped that further research here will improve our understanding of why they died out.

These particular animals are suspected to have been killed by a volcanic eruption and then swept to this point by floods. *Zhuchengtyrannus*' bones were discovered in 2009 purely by accident – by builders who were constructing a museum to house the finest fossils already found on the site.

As the name suggests, *Zhuchengtyrannus* was related to *Tyrannosaurus* and to judge by the few remains – some skull and jawbones – it was only a tad smaller. It would have been a gigantic killer: the lower jawbone is embedded with sharp, curved 10cm (4in) teeth, and likely prey would have included the 15m-long (49ft) hadrosaurs that form the bulk of the fossil remains at Zucheng. Like other tyrannosaurids it would have possessed tiny arms, two-fingered hands and powerful jaws capable of biting right through bone.

70mya

C

CARNIVOROUS

6000kg
(5.9 tons)

ZHUCHENG,
SHANDONG
PROVINCE,
CHINA

255

11m
(36ft)

LATEST
KNOWN
ARMOURED
DINOSAUR

(TARK-ee-ah)
TARCHIA GIGANTEA

CRETACEOUS

MAASTRICHTIAN	
CAMPANIAN	
SANTONIAN	LATE
CONIACIAN	
TURONIAN	
CENOMANIAN	
ALBIAN	
APTIAN	
BARREMIAN	EARLY
HAUTERIVIAN	
VALANGINIAN	
BERRIASIAN	

75–68
mya

H

HERBIVOROUS

4500kg
(4.4 tons)

MONGOLIA

Thanks to some perfectly preserved skeletons retaining skin impressions, we have a very clear image of this huge ankylosaurid, one of the biggest and latest to have lived. It had a heavily armoured skull covered with bony bumps and triangular spikes, and its back, sides and tail were similarly covered with bumps and scales and would have been equally impenetrable. This was vital to defend itself from predators such as *Tarbosaurus bataar*, the Asian tyrannosaur that occupied top spot in the food chain. The spikes and horns jutting from the rear of the skull might have served as a weapon when fighting rival members of the same species, as would the mighty club at the end of its tail. *Tarchia* lived in an arid desert environment and the fossils owe their survival to their immersion in sand dunes that over the aeons turned into stone. This was not only a very large ankylosaurid, it also had a particularly big braincase in proportion to body size – hence its name, which derives from the Mongolian for 'brainy'.

**8.5m
(28ft)**

257

CRETACEOUS

MAASTRICHTIAN
CAMPANIAN
SANTONIAN
CONIACIAN
TURONIAN
CENOMANIAN
ALBIAN
APTIAN
BARREMIAN
HAUTERIVIAN
VALANGINIAN
BERRIASIAN

LATE

EARLY

83–70 mya

H

HERBIVOROUS,
LOW BROWSER

40kg
(88lb)

CANADA

258

(STEG-oh-SEER-as)

STEGOCERAS VALIDUM

With a great bulge of bone atop its skull, a neck designed to absorb shocks and broad sturdy hips that provided a powerful base, it seems possible that *Stegoceras* spent a fair amount of its time head-butting other creatures. What is uncertain is which other creatures and why. For a long time it was thought that the males rutted with each other like stags, charging at one another and banging their heads to impress females and assert themselves as alpha *Stegoceras*. Now it is thought more likely that they butted each other from the side, perhaps as a way of warding off rivals and marking territory. None of the fossils found so far has borne any fractures or damage that suggests regular collision with other bony skulls, and *Stegoceras* was a hefty enough beast to suffer little lasting harm from regular butts to the flank. *Stegoceras* was a pachycephalosaurid and this genus, probably the best known within the family, was discovered in Canada in 1902 by Lawrence Lambe, who did much to establish his country as a centre for palaeontological research.

2.2m
(7ft)

(TIE-low-KEFF-ah-lee)

TYLOCEPHALE GILMOREI

MEANING 'SWOLLEN HEAD'

2m
(6ft 6in)

CRETACEOUS

MAASTRICHTIAN	
CAMPANIAN	LATE
SANTONIAN	
CONIACIAN	
TURONIAN	
CENOMANIAN	
ALBIAN	
APTIAN	
BARREMIAN	EARLY
HAUTERIVIAN	
VALANGINIAN	
BERRIASIAN	

75–72 mya

HERBIVOROUS

40kg (88lb)

MONGOLIA

Most pachycephalosaurids had heads topped with a dome of thick bone, and *Tylocephale*'s was the tallest. Compared with other bone-headed dinosaurs of the Late Cretaceous, its dome was narrow and unusually high at the back, with small spikes to the rear. With only one damaged skull known, it is hard for palaeontologists to determine its appearance but it was akin to *Stegoceras* and possibly even a species of that dinosaur rather than a genus in its own right. The Polish palaeontologists Halszka Osmólska and Teresa Maryańska described and named *Tylocephale*, which means 'swollen head', in 1974. From the location of its discovery in Mongolia's Gobi Desert (in the Late Cretaceous a habitat combining sand dunes and oases) experts surmised that pachycephalosaurids evolved in what is now Asia, moved to North America and then by *Tylocephale*'s time had returned to Asia.

MAASTRICHTIAN	
CAMPANIAN	
SANTONIAN	LATE
CONIACIAN	
TURONIAN	
CENOMANIAN	
ALBIAN	
APTIAN	
BARREMIAN	EARLY
HAUTERIVIAN	
VALANGINIAN	
BERRIASIAN	

81–75
mya

INSECTIVOROUS

3.5kg
(8lb)

OMNOGOV,
MONGOLIA

(shoo-VOO-ee-ah)
SHUVUUIA DESERTI

This little feathered creature is the only non-avian theropod known to have been able to move its snout region independently of its skull – an ability called prokinesis – which probably helped it gulp extra-large mouthfuls of wood-boring termites when it jabbed its jaws into the rotting trees that they inhabited. Like other alvarezsaurids its stunted arms ended in a hefty specialised claw that could have helped break into their nests. In *Shuvuuia* this was joined by two minuscule additional claws. As a relatively small and slight dinosaur it seems possible it could have subsisted on such meagre creatures.

Its name is the Mongolian for 'bird', and at times palaeontologists have leaned towards classifying it as one, but at present the consensus is that it was a bird-like dinosaur. While analysis of the fossils has found beta keratin traces, a sure sign of feathers, there is no way that it could have flown – so it developed long, slender legs that made it a fast runner, its only method of escaping predators.

1m
(3ft)

(STROOTH-ee-oh-MEE-mus)

STRUTHIOMIMUS ALTUS

CRETACEOUS

MAASTRICHTIAN	
CAMPANIAN	LATE
SANTONIAN	
CONIACIAN	
TURONIAN	
CENOMANIAN	
ALBIAN	
APTIAN	
BARREMIAN	EARLY
HAUTERIVIAN	
VALANGINIAN	
BERRIASIAN	

Other species:
S. sedens

4m
(13ft)

76–74
mya

OMNIVOROUS

150kg
(330lb)

ALBERTA,
CANADA

With its small head, beak, long neck and legs and almost certainly feathered body, it is easy to see why this dinosaur was given a name meaning 'ostrich-like' – superficially at least, the most significant difference between the animals' appearance is *Struthiomimus*' long horizontal tail. It was the first of the ornithomimids to be discovered from a complete skeleton, which gave a great insight into this family, the fastest dinosaurs known to have lived. *Struthiomimus* ran at around 30 to 40mph, and possibly up to 50mph in short bursts despite standing only around 1.4m (4ft 7in) tall. The long and robust tail gave it great balance, enabling it to attain such velocity.

It lived in the forested coastal floodplains and swamps of what is now Canada, where its speed would have helped it to evade predators such as *Albertosaurus* and the fearsomely clawed dromaeosaurids of the Late Cretaceous. There is some debate over whether *Struthiomimus* was itself a predator. Some fossils have been found to contain 'stomach stones', the lumps of grit and gravel swallowed by herbivorous birds to this day to grind up plant matter and aid digestion. But the sharp beak suggests that it might well have been an omnivore, perhaps eating small animals such as lizards and insects.

261

MYTHOLOGY

Enormous ancient skeletons of strange animals were known to humans long before we had any concept of dinosaurs. Naturally enough people's innate curiosity and need to make sense of the world led them to develop all manner of theories and folklore, and some of these ideas remain very familiar to us today.

One example is the griffin, that strange amalgam of an eagle's head and wings and a lion's body. It was first mentioned in 675BCE by a Greek writer and explorer named Aristeas of Proconnesus, who had encountered a group of Scythian nomads prospecting for gold between the Altai and Tien Shan mountain ranges at the borders of modern China, Mongolia, Siberia and Kazakhstan. Aristeas recorded their tales of grotesque, ferocious creatures that guarded hoards of gold in the wilderness, warding off trespassers with their savage beaks and claws; and he repeated the Scythians' assurances that their brave ancestors had killed many of these beasts and left their bones in the desert for all to see. The legend flourished throughout classical antiquity but in 2000 American folklorist Adrienne Mayor noted that the mythical creature strongly resembles a real animal whose remains are prevalent in the Gobi Desert: not a lion, or an eagle, but a *Protoceratops*. The small herbivorous dinosaur's bones and eggs have lain scattered across this region for millions of years, fossilised into a cream-coloured stone that is easily visible against the rosy hues of the Djadochta Formation rocks. With its beaked skull, wing-like neck frill and hefty clawed feet, the *Protoceratops'* skeleton bears an uncanny similarity to the griffin.

Nor is it any coincidence that dragons are so significant in Chinese mythology, given that country's wealth of dinosaur fossils. Reptilian skulls comprising horned snouts and savagely-toothed jaws, massive bodies with powerful limbs, a long serpentine tail... all that differentiates many dinosaurs' skeletons from the archetypal dragon is the lack of wings and fiery breath. A Chinese document from the second century BCE records that many such bones were found during the digging of a canal, which was therefore named the Dragon-Head Waterway. In the 1950s many of the first identified Triassic-era fossils in China were found at a mountain in Guizhou province traditionally known as Lurking Dragon Hill, which bore abundant exquisite remains of a

30cm-long (12in) long marine reptile now termed *Keichousaurus*. For centuries Chinese people had collected these little 'dragons' as good luck charms, some even deliberately using the fossiliferous rocks when building their houses. The country is now at the forefront of palaeontological research but in places the old folklore persists. In 2007 it was reported that villagers in Ruyang County had excavated several tons of fossil remains of a large sauropod and crushed them into powder to use in medicinal soup, thinking that they were dragon bones and would bring good health and fortune. The cultural history is also reflected in Chinese palaeontologists' common practice of using 'long', the word for 'dragon', when scientifically naming dinosaurs: for instance *Mei long*, the 'sleeping dragon', and *Guanlong*, the 'crowned dragon'.

Many similar but less familiar examples occur in folklore around the world. The Himalayan region of northern India is rich with giant petrified bones and has its own dragon folklore. The Tuareg nomads of Niger spoke of a monster they called Jobar, whose huge bones lay scattered around the desert. When they found Chicago palaeontologist Paul Sereno excavating some relatively paltry remains nearby in 1997, the Tuareg directed him toward the far bigger 'Jobar' bones and he identified them as the remains of a large sauropod, which he and colleagues later named *Jobaria* (page 76).

Native Americans have a wealth of stories that explain the presence of massive fossils in their territories. The Sioux tribe believed that a horned, four-legged water serpent called Unktehi lived in ancient times until it was destroyed by flying beasts they called Thunderbirds; these two myths may have been inspired by dinosaur and pterosaur fossils respectively.

While it is easy to take a patronising attitude towards such ideas, so much of what we term mythology originally springs from a laudable urge to interpret and explain the world from the limited available evidence. It is also evidence of these fascinating relics' power to make people marvel and speculate, even before modern scientists developed a clear idea of the animals that left them behind.

MAASTRICHTIAN	
CAMPANIAN	
SANTONIAN	LATE
CONIACIAN	
TURONIAN	
CENOMANIAN	
ALBIAN	
APTIAN	
BARREMIAN	EARLY
HAUTERIVIAN	
VALANGINIAN	
BERRIASIAN	

82-74
mya

H

HERBIVOROUS

3000kg
(2.9 tons)

ALBERTA,
CANADA, AND
MONTANA, USA

264

(GRIP-oh-SORE-us)

GRYPOSAURUS NOTABILIS

MEANING
'HOOK-
NOSED
LIZARD'

Its distinctive 'Roman nose' sets this hadrosaur apart from other genera, but it shared with them an insatiable appetite for vegetation, which it browsed from bushes and low trees. This pronounced curve may have distinguished between males and females, and perhaps functioned as a weapon for butting rivals. Inflatable sacs may have been attached that it could use for display. Several excellent fossils are known from Montana in the USA and Alberta, Canada, including one with preserved skin, which shows that *Gryposaurus* had three types of scales: little pyramids 3.8cm (1.5in) high along its flank and tail, regular scales on its neck and sides, and a row of triangular scales along its spine. Lawrence Lambe described *Gryposaurus*, whose name means 'hook-nosed lizard', after the first specimen's discovery in 1913 in Alberta.

8m
(26ft)

(LIN-he-NIKE-us)
LINHENYKUS MONODACTYLUS

60cm
(2ft)

CRETACEOUS

MAASTRICHTIAN	
CAMPANIAN	
SANTONIAN	LATE
CONIACIAN	
TURONIAN	
CENOMANIAN	
ALBIAN	
APTIAN	
BARREMIAN	EARLY
HAUTERIVIAN	
VALANGINIAN	
BERRIASIAN	

84–75
mya

INSECTIVOROUS

450g
(1lb)

LINHE,
INNER
MONGOLIA,
CHINA

265

This tiny alvarezsaurid was only the size of a parrot and took the family's characteristic diminished digits to an extreme unknown prior to its discovery in 2011. *Mononykus'* (page 287) name suggests it had only one finger but actually it had the vestiges of two other fingerbones. *Linhenykus*, however, did not even have these vestiges, so it may fairly be called the first truly one-fingered dinosaur. Oddly enough this doesn't represent the culmination of the alvarezsaurids' strange adaptation, as *Linhenykus* was one of the earlier members of the family. The single finger was actually smaller and weaker than in other genera, though *Linhenykus* had a strong chest and arms. Like the others, however, its hands probably evolved to enable it to break into anthills and termite galleries. The Chinese palaeontologist Xu Xing and colleagues named it after the city of Linhe, Inner Mongolia, where its fossil was found.

CRETACEOUS

MAASTRICHTIAN	
CAMPANIAN	
SANTONIAN	LATE
CONIACIAN	
TURONIAN	
CENOMANIAN	
ALBIAN	
APTIAN	
BARREMIAN	EARLY
HAUTERIVIAN	
VALANGINIAN	
BERRIASIAN	

76–73 mya

HERBIVOROUS

2500kg
(2.4 tons)

WESTERN
NORTH AMERICA
FROM CANADA
DOWN TO
NEW MEXICO

266

(PAR-a-sore-OL-o-fuss)

PARASAUROLOPHUS WALKERI

Other species: P. tubicen, P. cyrtocristatus

Just as we are finally learning some of the dinosaurs' colours (page 208), we are also getting hints of the sounds they made. The slender, 1.2m-long (4ft) tubular crest protruding backwards from *Parasaurolophus*' nostrils has been much discussed since this lambeosaur was named in 1922. Inaccurate suggestions include that *Parasaurolophus* was amphibious and the crest served as a snorkel, or as a kind of oxygen tank for long periods underwater. Others said it was a weapon or contained tissue that heightened its sense of smell. But these ideas have all been dismissed – there was no hole at the top of the crest so it couldn't have been a snorkel, nor could it have held enough air to be of much use to such a large creature. Further research showed that its olfactory nerves were elsewhere within the skull.

Instead it seems that this most distinctive feature had the purposes of both visual and aural communication. In the mid-1990s scientists from the New Mexico Museum of Natural History and Science took a recently discovered near-complete skull to a hospital in Albuquerque and carried out 350 computed tomography (CT) scans. With this data they recreated a computer model of the skull and then subjected it to virtual 'air pressure'. The tests showed that the twisting tubes within the crest would have produced a low, mournful note loud enough to resonate across the plains and through the jungle. *Parasaurolophus* would have been able to control and fluctuate the note, enabling more sophisticated communication, and it is thought that each animal had its own distinctive voice. You can hear recreations of *Parasaurolophus*' voice online, for instance by searching on YouTube.

9m
(30ft)

The other likely purpose was for display – there may have been a flap of skin between its crest and neck, which could have been used for signalling. Finally the crest may have helped regulate temperature, by soaking up heat during the day and gradually releasing it at night.

Like other duck-billed dinosaurs *Parasaurolophus* had hundreds of small teeth, known as a dental battery, enabling it to chew through masses of vegetation. It was primarily four-legged but could run on its hind legs to escape predators. The three known species had differently curved crests, and it is thought that in each case males probably had bigger crests than females: in fact the variety with the shortest crest, *P. cyrtocristatus*, may actually have been a female or juvenile of one of the other two species.

MAASTRICHTIAN	
CAMPANIAN	
SANTONIAN	LATE
CONIACIAN	
TURONIAN	
CENOMANIAN	
ALBIAN	
APTIAN	
BARREMIAN	EARLY
HAUTERIVIAN	
VALANGINIAN	
BERRIASIAN	

80–70
mya

HERBIVOROUS

45kg
(100lb)

MONGOLIA

268

(HOME-ah-lo-KEFF-ah-lee)
HOMALOCEPHALE CALATHOCERCOS

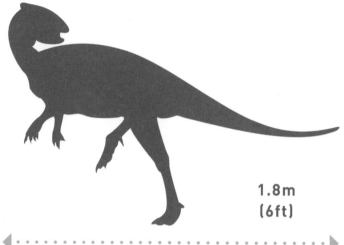

**1.8m
(6ft)**

This primitive pachycephalosaur had a thin skull compared with better known members of the 'bonehead' family: its forehead was only twice as thick as most other bird-hipped dinosaurs', whereas its more advanced relatives' skulls were 20 times as thick. Its unusually wide hips cause some experts to think that *Homalocephale* gave birth to live young rather than laying eggs, but others believe the purpose was to protect its internal organs when rivals butted its flanks. Its long legs suggest a fast runner. However, there is a note of doubt about *Homalocephale*, whose name means 'level head': some palaeontologists suggest that the single known skeleton was a juvenile *Stegoceras*.

(gee-GANT-oh-rap-ter)
GIGANTORAPTOR ERLIANENSIS

MAASTRICHTIAN	
CAMPANIAN	
SANTONIAN	LATE
CONIACIAN	
TURONIAN	
CENOMANIAN	
ALBIAN	
APTIAN	
BARREMIAN	
HAUTERIVIAN	EARLY
VALANGINIAN	
BERRIASIAN	

One of the most monstrous-looking feathered creatures ever to exist, *Gigantoraptor* stood the height of a giraffe and possessed 20cm (8in) slashing claws. At the top of its powerful ostrich-like neck was a toothless beaked skull more than half a metre (1½ft)long. Its arms may well have sported display feathers like its fellow oviraptorosaurs – but it was an incredible 35 times bigger than *Citipati*, the next largest related genus.

Gigantoraptor emerged when a team of Japanese documentary filmmakers accompanied Xu Xing to a region of the Gobi Desert known for its sauropod bones. He found a thighbone and soon realised that it was unlike anything seen before. Much of the skeleton followed. On announcing the discovery he said: 'If you found a mouse as big as a pig you would be very surprised – it was the same when we found the *Gigantoraptor*.'

But despite its fearsome appearance, this was probably not the nightmarish predator it appears. It was not a true 'raptor', such as *Velociraptor* (page 282), but a relative of the probably omnivorous *Oviraptor* (page 252). *Gigantoraptor* most likely ate plants, and perhaps eggs and molluscs.

No feather impressions were found with the fossil but Xu and his colleagues believe it shared its relatives' feather covering to some extent. Even if it did not need a covering of downy feathers for insulation, as smaller oviraptorosaurs did, it probably retained arm feathers for display and covering eggs while brooding. Despite its bird-like features, it was not on the evolutionary line that led to modern birds, but its discovery did solve one mystery. For years experts had wondered what creature could have laid the 3m-wide (10ft) rings of massive, 53cm-long (21in) fossilised eggs found in the Gobi Desert. In *Gigantoraptor* they had their answer.

8m (26ft)

70mya

OMNIVOROUS

1400kg
(1.3 tons)

MONGOLIA

CHASMOSAURUS

Chasmosaurus (pictured) was a long-frilled ceratopsian, which shared a habitat with short-frilled genera such as *Centrosaurus*.

CRETACEOUS

MAASTRICHTIAN	
CAMPANIAN	
SANTONIAN	LATE
CONIACIAN	
TURONIAN	
CENOMANIAN	
ALBIAN	
APTIAN	
BARREMIAN	EARLY
HAUTERIVIAN	
VALANGINIAN	
BERRIASIAN	

(SENT-ro-SORE-us)

CENTROSAURUS APERTUS ·············▶

One stormy day around 76mya a huge herd of *Centrosaurus* were ambling their way over the coastal plains of what's now western Canada when the skies darkened, the winds got up to gale force and rain began to fall... and it fell long and hard until a vast swathe of lowland lay submerged under floodwater. Birds flew away, small mammals and reptiles scurried up trees, but the massive ceratopsids had no high ground nearby to run to. Hundreds of them drowned. When the waters receded their bodies settled in clusters on the ground and were picked over by carnivores before being enclosed in the mud. Now their remains are preserved in one of the world's great dinosaur graveyards, the 'Hilda mega-bonebed', named after the nearest small town in Canada's Alberta province. The mud turned into mudstone, which was exposed again around 12,000 years ago when a glacier gouged its way through the rock, leaving behind it a valley. Now the South Saskatchewan River runs through the valley, and its waters often erode fossilised bones from the valley walls.

Covering an estimated 0.9sq miles, the bonebed was identified in 1959 but it was not until 2010, after a decade's intensive analysis, that its contents were properly described.

After excavating a series of sample areas and counting the number of fossils in each, palaeontologists multiplied the number to obtain an idea of how many *Centrosaurus'* remains lie within the rock stratum, much of which is inaccessible. Their calculations produced the estimated figure of 667, which shows that *Centrosaurus* herds – and presumably those of other ceratopsians – were far larger than previously known, containing hundreds and possibly thousands of these bulky, 6m-long (20ft) herbivores.

Centrosaurus had a large horn on its snout, two smaller horns above its eyes and yet more on its frilled neck shield, which had two openings designed to reduce its weight. It is is classified as a short-frilled ceratopsian, as opposed to long-frilled genera such as *Chasmosaurus*, with which it shared its habitat. The largest horn faced backwards, forwards or straight upwards in different individuals, and it is not certain why. It would have used its spikes and shield to defend against predators such as *Daspletosaurus*, but evidently it could do little to cope with the floods that periodically devastated North America's coastal flatlands in the Cretaceous.

6m (20ft)

76–74 mya

HERBIVOROUS, LOW BROWSER

2000kg (1.9 tons)

ALBERTA, CANADA

CRETACEOUS

MAASTRICHTIAN	
CAMPANIAN	
SANTONIAN	LATE
CONIACIAN	
TURONIAN	
CENOMANIAN	
ALBIAN	
APTIAN	
BARREMIAN	EARLY
HAUTERIVIAN	
VALANGINIAN	
BERRIASIAN	

70mya

HERBIVOROUS

5000kg
(4.9 tons)

MONGOLIA

272

(THER-ih-zeen-oh-SORE-us)
THERIZINOSAURUS CHELONIFORMIS

Theropods were vicious carnivores or bird-like omnivores – weren't they? These bizarre and mysterious creatures undermine that notion. *Therizinosaurus* has intrigued scientists ever since its discovery in Mongolia in 1948. The first incomplete remains – some flattened ribs and mighty arms and claws – were thought to come from a turtle-like lizard. In 1954 a Russian palaeontologist named Evgeny Maleev named it *Therizinosaurus cheloniformis*, which means 'scythe lizard, turtle-shaped'. But as other discoveries emerged the picture grew fuller – and stranger. More excavations in the 1950s unearthed further claws and limbs, and palaeontologists became convinced that

10m
(33ft)

CRETACEOUS

MAASTRICHTIAN	
CAMPANIAN	
SANTONIAN	LATE
CONIACIAN	
TURONIAN	
CENOMANIAN	
ALBIAN	
APTIAN	
BARREMIAN	EARLY
HAUTERIVIAN	
VALANGINIAN	
BERRIASIAN	

AND ..... *SEGNOSAURUS GALBINENSIS*

(SEG-no-SORE-us)

this was a dinosaur. In 1970 another Russian named Anatole Rozhdestvensky argued that it was a large theropod that lived on ants, using its claws to break into anthills. He was right that it was a theropod, but misguided in the belief that something the size of *Tyrannosaurus* could subsist on such tiny creatures.

The sickle-shaped claws alone measured up to 90cm (3ft), and the arms were 2m (6ft 6in) long. With most theropods such deadly equipment would be used to kill prey, and sure enough *Therizinosaurus* was still imagined to be a carnivore. However, its enormous bulk would have prevented it from pursuing land animals effectively, so scientists speculated that it ate fish instead. Working from minimal finds that bore little relation to other known dinosaurs meant that there was a large amount of guesswork involved in reconstructing *Therizinosaurus'* appearance. Some imagined it with a carnosaur's skull and physique, perhaps with a killing claw on the foot like *Deinonychus*.

Then the discovery of the smaller and much older *Segnosaurus* in 1973 shed new light. *Segnosaurus* had scaled-down versions of *Therizinosaurus'* arms and claws, but the fossil remains also included skull fragments, which revealed a downturned jaw and leaf-shaped teeth. So *Segnosaurus* was herbivorous, one of the few plant-eating theropods, and *Therizinosaurus* almost certainly was too.

Their claws were most likely used for pulling down foliage from trees, but they must also have been useful as defence against predators such as *Tarbosaurus*, the great tyrannosaur of Late Cretaceous Mongolia. Why this branch of the theropod family evolved into herbivores is a question that palaeontologists would still love to answer.

90mya

HERBIVOROUS

1300kg (1.2 tons)

MONGOLIA

6m (20ft)

MAASTRICHTIAN	
CAMPANIAN	
SANTONIAN	LATE
CONIACIAN	
TURONIAN	
CENOMANIAN	
ALBIAN	
APTIAN	
BARREMIAN	EARLY
HAUTERIVIAN	
VALANGINIAN	
BERRIASIAN	

75–65
mya

OMNIVOROUS

50kg
(110lb)

WESTERN
NORTH
AMERICA

274

(TROH-oh-don)

TROODON FORMOSUS

A single serrated tooth was all the early American palaeontologist Joseph Leidy had to work with when he described *Troodon* in 1856, hence its name meaning 'wounding tooth'. He thought it was a lizard but subsequent work showed *Troodon* to be a dinosaur – the most intelligent ever discovered. It had a brain six times heavier than other similar-sized dinosaurs, much like that of a modern emu. It had extremely acute hearing owing to an enlarged middle-ear cavity and the fact that one of its ears was a touch higher than the other, a trait otherwise only known in owls. Its exceptionally large, forward-facing eyes gave it binocular vision even in dark conditions, perfect for tracking its prey at night. It may have attacked young duck-billed dinosaurs as they slept: some juvenile *Edmontosaurus* fossils show *Troodon* bite-marks.

Troodon was a long-legged fast runner with a retractable killing claw on each foot, and was probably feathered. It gives its name to the family of troodontids, small theropods that occupy a niche between the horrifically clawed dromaeosaurids and the ostrich-like ornithomimids. *Troodon* specimens have been found in Montana, Wyoming, Alberta in Canada, and even as far north as Alaska. Interestingly

2m
(6ft 6in)

the Alaskan fossils suggest that it grew bigger there, perhaps up to 4m (13ft) long rather than 2m (6ft 6in) elsewhere. This is attributed to environmental factors: it has been known since the mid-19th century that animals living at high latitudes tend to grow larger than comparable genera nearer the equator. One proposed explanation is that plants growing in cool climates contain more nitrogen, making them more nutritious. This helped herbivores to grow bigger, and thus in turn the theropods that hunted them.

BOROGOVIA GRACILICRUS

CRETACEOUS

MAASTRICHTIAN
CAMPANIAN
SANTONIAN
CONIACIAN
TURONIAN
CENOMANIAN
ALBIAN
APTIAN
BARREMIAN
HAUTERIVIAN
VALANGINIAN
BERRIASIAN

LATE

EARLY

70–65 mya

C

CARNIVOROUS

13kg (25lb)

MONGOLIA

**2m
(6ft 6in)**

*'Twas brillig, and the slithy toves
Did gyre and gimble in the wabe:
All mimsy were the borogoves,
And the mome raths outgrabe.*

So begins Lewis Carroll's poem 'Jabberwocky', which appears in *Through the Looking Glass* and is an acknowledged classic of nonsense verse, populated by a menagerie of curious beasts that Carroll drew from the depths of his brilliant imagination. Later in the book a confused Alice receives a translation of sorts from Humpty Dumpty, who explains that 'a "borogove" is a thin shabby-looking bird with its feathers sticking out all round – something like a live mop'.

When the Polish palaeontologist Halszka Osmólska needed a name for a small feathered troodont discovered in the Gobi Desert she decided to make the borogove a reality. *Borogovia*, a bird-like stealth hunter with a beak full of small sharp teeth perfect for eating small lizards and mammals, is only known from its hind limbs but appears to have been similar to *Saurornithoides* – only of thinner build, as befits the name, and with a straighter killing claw on its second toe. Whether its feathers stuck out all round as Humpty suggests will only be confirmed by a far better quality fossil.

275

CRETACEOUS

MAASTRICHTIAN	
CAMPANIAN	
SANTONIAN	LATE
CONIACIAN	
TURONIAN	
CENOMANIAN	
ALBIAN	
APTIAN	
BARREMIAN	EARLY
HAUTERIVIAN	
VALANGINIAN	
BERRIASIAN	

76.4–75.5
mya

HERBIVOROUS

4000kg
(3.9 tons)

UTAH,
USA

276

(YOO-tah-SERR-a-tops)

UTAHCERATOPS GETTYI · · · · · · · · AND · · ·

When researchers unveiled *Kosmoceratops* in 2010 and proclaimed it the most ornately decorated dinosaur ever discovered, it rekindled interest in one of the last uncharted terrains in American palaeontology.

The Western Interior Seaway split the USA into two continents for 27m years during the Cretaceous: to its west lay Laramidia, a slender strip running from modern Alaska to Mexico, and the larger landmass to the east is termed Appalachia. Every new discovery made in Laramidia makes it more apparent that this was one of the Cretaceous world's great dinosaur hotspots, and the discovery of *Kosmoceratops* and its contemporary *Utahceratops* emphasised this. Both were a kind of ceratopsian termed a chasmosaur, and both left fossils found in the Grand Staircase-Escalante National Monument, Utah. The 1.9-million-acre region is so-called because it comprises

a series of cliffs and plateaus that combine to look like an immense flight of steps leading down to the Grand Canyon. Around 75mya it was a lush, swampy patch of southern Laramidia beside that dividing stretch of saltwater. Now it is an arid and inhospitable landscape considered one of the last mysteries in American dinosaur research. To find *Kosmoceratops* and *Utahceratops*, palaeontologists from the Utah Museum of Natural History had to hike many miles in unforgiving conditions and then commission a helicopter to carry the bones away. But it was worth it. *Kosmoceratops* garnered more attention when it was announced owing to its bizarre features. Its massive 2m-long (6ft 6in) skull bore 15 horns: one above its nose and each eye, one on each cheekbone, and a row of ten small spikes on its forehead, eight of them pointing forwards like a fringe of bones.

7m
(23ft)

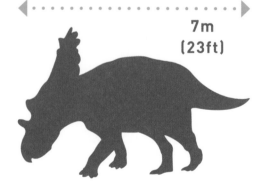

CRETACEOUS

MAASTRICHTIAN	
CAMPANIAN	
SANTONIAN	LATE
CONIACIAN	
TURONIAN	
CENOMANIAN	
ALBIAN	
APTIAN	
BARREMIAN	EARLY
HAUTERIVIAN	
VALANGINIAN	
BERRIASIAN	

(KOS-mo-SERR-a-tops)

KOSMOCERATOPS RICHARDSONI

5m (16ft)

76.4–75.5 mya

H

HERBIVOROUS

2500kg (2.4 tons)

UTAH, USA

Utahceratops stood 50cm taller at 2m (6ft 6in) high but was less impressively decorated. Both were important, however. First, they left near-complete skeletons that aided understanding of ceratopsian anatomy. Second, they heightened the concentration of known dinosaur species within a relatively small area. Dr Scott Sampson, lead author of the journal article describing the finds, said: 'Today we have a handful of rhino-to-elephant sized mammals living in Africa. At present, it seems that there were at least 15 to 20 rhino-to-elephant sized animals living on Laramidia 76m years ago, despite the fact that it was less than one-fifth the size of Africa.'

The third reason for their significance was their pronounced anatomical differences from horned dinosaurs found in the northern half of Laramidia, which reinforced palaeontologists' belief that the continent had two distinct evolutionary centres. Fourth, in *Kosmoceratops'* case particularly, they improve understanding of the horned dinosaurs' extravagant ornamented skulls. 'Most of these bizarre features would have made lousy weapons to fend off predators. It's far more likely that they were used to intimidate or do battle with rivals of the same sex, as well as attract individuals of the opposite sex,' said Dr Sampson, likening them to deer antlers.

KOSMOCERATOPS

Extravagantly ornamented *Kosmoceratops* tussle with one another as a group of the feathered troodontid *Talos* look on.

CRETACEOUS

MAASTRICHTIAN	
CAMPANIAN	
SANTONIAN	LATE
CONIACIAN	
TURONIAN	
CENOMANIAN	
ALBIAN	
APTIAN	
BARREMIAN	EARLY
HAUTERIVIAN	
VALANGINIAN	
BERRIASIAN	

77–76 mya

H

HERBIVOROUS

3000kg
(2.9 tons)

MONTANA,
USA

280

(MY-ah-SORE-ah)
MAIASAURA PEEBLESORUM

MEANING 'GOOD MOTHER LIZARD'

9m
(30ft)

This huge hadrosaur had a wide, duck-like beak and a crested head, but is best known for its maternal instincts. American palaeontologists Jack Horner and Bob Makela named the 'good mother lizard' in 1979 after its remains were found at a location dubbed 'Egg Mountain', a plateau in the Rockies of Montana littered with fossilised dinosaur nests (see page 134). As well as the broken remains of grapefruit-sized eggs, the site held the fossils of juvenile *Maiasaura*. Juvenile fossils have since been found at other dinosaurs' breeding colony sites, but those of *Maiasaura* were twice the age of any others known, suggesting that *Maiasaura* cared for its young for an unusually long period. Between returning to its nesting sites, *Maiasaura* traversed the plains of the western USA in herds perhaps 10,000-strong, feeding on leaves and berries. If you're wondering why its name ends in 'saura' rather than the far more common 'saurus', the former is feminine and the latter is masculine, and it was assumed that most of the adult fossils found by the nests were female. Given that most dinosaurs' fossils cannot be attributed to either sex, it would seem fairer if more names were given the feminine form. *Maiasaura* also has the distinction of being the first dinosaur in space. In 1985 a Montana-born astronaut named Loren Acton brought along a *Maiasaura* fossil from his home state when he took part in a space shuttle mission.

(ah-BELL-ee-SORE-us)

ABELISAURUS COMAHUENSIS ·········▸

Known only from a single monstrous skull, this carnivore gives its name to the family of devastating predators that sprinted and pounced their way around the southern hemisphere in the Jurassic and Cretaceous. Dinosaurs that we now term abelisaurs had been known since the early 1900s but it was the discovery of *Abelisaurus* and *Carnotaurus* in the early 1980s that linked them all together. Features in common include smaller teeth than most theropods, areas of corrugated bone on their skulls, powerfully muscled necks and minuscule arms. *Abelisaurus* is hard to describe given the paucity of remains, but palaeontologists suspect it was a primitive member of the family. Its 83cm-long (33in) skull was incomplete but clearly contained large gaps known as fenestrae (as in fenestration, meaning windows). These open spaces reduced the head's weight, allowing it to possess huge jaws full of short sharp teeth without being so heavy as to create an imbalance. The genus was named for Robert Abel, former director of the Argentine museum in which the skull is displayed, and the specific name refers to the fossil site's location in Comahue, Patagonia.

CRETACEOUS

MAASTRICHTIAN	
CAMPANIAN	
SANTONIAN	LATE
CONIACIAN	
TURONIAN	
CENOMANIAN	
ALBIAN	
APTIAN	
BARREMIAN	EARLY
HAUTERIVIAN	
VALANGINIAN	
BERRIASIAN	

83–78 mya

CARNIVOROUS

2000kg (1.9 tons)

ARGENTINA

6.5m
(21ft)

◂ ·········· ▸

CRETACEOUS

MAASTRICHTIAN	
CAMPANIAN	
SANTONIAN	LATE
CONIACIAN	
TURONIAN	
CENOMANIAN	
ALBIAN	
APTIAN	
BARREMIAN	EARLY
HAUTERIVIAN	
VALANGINIAN	
BERRIASIAN	

83–70 mya

CARNIVOROUS

15kg
(33lb)

MONGOLIA

282

(ve-LOSS-ih-rap-tor)
VELOCIRAPTOR MONGOLIENSIS

Picture the scene: a herd of *Protoceratops* lies asleep out in the Mongolian desert, the landscape glowing silver in the light of a great pale moon. All appears still... but that is an illusion, for stealthily approaching through the night comes a small predator with eyes only for a sleeping herbivore. A skilful double-pronged barrage of claws and jaws ensues and *Velociraptor* has its victim. This was one of the most effective little hunters of the Late Cretaceous, the first of the dromaeosaurids to be identified, and we know a good deal about its appearance and how it lived.

First of all, disregard the image you may have from *Jurassic Park* – *Velociraptor* was small and feathered, rather than large and scaly. But while *Velociraptor* couldn't have deduced how to turn a door handle, it did sit towards the sharper end of the dinosaurs' intelligence scale. Professor Lawrence Witmer, a dinosaur brain specialist, equates its intelligence to that of a modern bird of prey. It probably ate anything it could, perhaps even its own kind; one specimen's bones bear bite-marks fitting *Velociraptor* teeth. We know that it hunted *Protoceratops* owing to one of the most amazing fossils ever found: the 'Fighting Dinosaurs' recovered from Mongolia in 1971, which preserves the two animals' skeletons locked in a lethal grapple. The *Velociraptor* has its left foot's 'killing claw' jabbed into the *Protoceratops*' neck while its left hand slashes at its face. The herbivore is defending itself by clamping its tough beak hard around its attacker's right arm. It's thought that they killed each other and were then quickly engulfed in a sand drift.

2.5m
(8ft)

We also have good data indicating that *Velociraptor* was nocturnal. Analysis of its sclerotic rings (the circle of bone around the eye connected to a muscle that controlled how much light entered the pupil) shows that it could see well in the dark.

Finally we know that it bore a fine array of true feathers, the sort you'd recognise on a bird today, owing to the discovery in 2007 of a fossil arm bone studded with small indentations where the quills joined the bone (see page 97). Information about *Velociraptor* has

been gathered since 1923, when researchers from the American Museum of Natural History travelled to the Gobi Desert to trace the origins of human life. They failed on that account but found many fascinating animals' fossils, including a crushed skull and a characteristic sickle claw. A year later museum president Henry Fairfield Osborn coined one of the most memorable dinosaur names: it means 'fast hunter' and conjures very well the dynamism of this aggressive bird-like predator.

OUR CHANGING CONTINENTS

LATE TRIASSIC

During the Earth's history the continents have been united into one vast landmass, drifted apart and reunited several times over. In the Triassic, they were connected in the form of a super-continent known as Pangaea (meaning 'entire Earth'). South America and Africa slotted together to form the bulk of the southern part, referred to as Gondwana, along with Antarctica, India and Australia. The land north of the equator is termed Laurasia, which combined North America, Europe and Asia (apart from India). Surrounding them all was Panthalassa ('all ocean'), which included the Pacific to the west and the Tethys Ocean to the east. Pangaea existed from approximately 300mya to 200mya.

LATE JURASSIC

The Jurassic saw Pangaea's constituent parts begin to rift. Gondwana and Laurasia separated, with the Pacific and the Tethys connecting to form a separating sea between these northern and southern landmasses. Modern South America and Africa began to divide, while North America and Europe moved apart: what is now Britain lies north of a string of islands that are today connected as mainland Europe. Antarctica and Australia remained united at this stage. The fracturing of the land into smaller continents and islands led to a more humid climate in which the dinosaurs flourished.

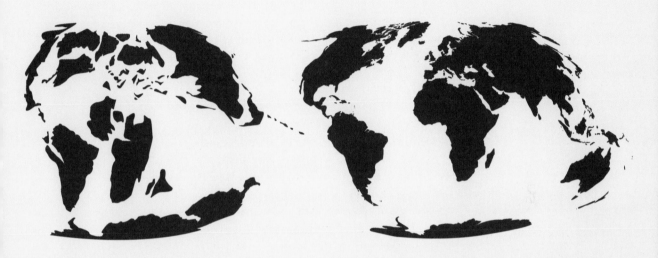

LATE CRETACEOUS

By the Late Cretaceous, the world began
to assume an appearance recognisable to
us today. Australia had almost sheared
away from Antarctica, while Africa and
South America were completely separate,
with northwestern Africa forming an island
distinct from the rest of the continent.
Madagascar lay off Africa's east coast,
and beside it India was drifting north
towards Asia, where it now lies. North
America was divided into two areas by the
Western Interior Seaway: Appalachia lay
to the east, and the slender strip known
as Laramidia was home to a diversity
of dinosaurs including tyrannosaurs,
ceratopsians, hadrosaurs and sauropods.

THE MODERN WORLD

The world map we are all familiar with,
showing the continents as they appear
today.

MAASTRICHTIAN	
CAMPANIAN	
SANTONIAN	LATE
CONIACIAN	
TURONIAN	
CENOMANIAN	
ALBIAN	
APTIAN	
BARREMIAN	EARLY
HAUTERIVIAN	
VALANGINIAN	
BERRIASIAN	

74–70 mya

OMNIVOROUS

170kg (370lb)

ALBERTA, CANADA

(dro-MISS-ee-oh-mee-mus)

DROMICEIOMIMUS SAMUELI

MEANING 'EMU MIMIC'

Is this the fastest-running dinosaur known? It's calculated that its long muscular legs could propel *Dromiceiomimus* along at up to 50mph, making it an elusive target for its larger fellow theropods. Skeletons suited to sprinting have longer lower legs than upper legs, and no other ornithomimid had shins as proportionally long as *Dromiceiomimus*. While it was almost certainly omnivorous, it probably didn't use its speed for chasing prey. Its slender, ostrich-like beak suited a diet of plant matter and very small animals: lizards, tiny mammals, perhaps insects. Its large eyes suggest it caught them in low light conditions. The name means 'emu mimic' and like other ornithomimids it must have looked very like an ostrich or emu, only with a large tail and long arms that had slender, tapering fingers. However, the ornithomimid dinosaurs have proved confusing to classify, and some palaeontologists now think that *Dromiceiomimus* should be included as a species of *Ornithomimus*.

3.5m (11ft)

(mon-oh-NIKE-us)
MONONYKUS OLECRANUS

CRETACEOUS

MAASTRICHTIAN	
CAMPANIAN	
SANTONIAN	LATE
CONIACIAN	
TURONIAN	
CENOMANIAN	
ALBIAN	
APTIAN	
BARREMIAN	EARLY
HAUTERIVIAN	
VALANGINIAN	
BERRIASIAN	

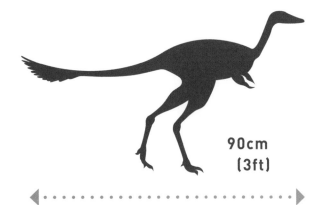

**90cm
(3ft)**

With hands reduced to a single claw and minuscule second and third digits, this is one of the strangest and most specialised of the advanced theropods. Its unveiling in 1993 prompted palaeontologists to furrow their brows: what were those bizarre hands adapted for? Palaeontologist Phil Senter's close study of the arm bones, published in 2005, showed that they could not have been used for burrowing as some suspected. This made sense, as *Mononykus'* long neck, legs and tail would appear a hindrance to such a lifestyle and more suited to fast running across the Mongolian desert; contrast its lithe form with that of a typical burrowing animal such as a mole. Instead Senter's work revealed that the arms and hands had just the same movement range as anteaters and pangolins – so it is surmised that *Mononykus* and its fellow alvarezsaurids were the anteaters of the Late Cretaceous.

At first *Mononykus* was considered a flightless bird, but then it was argued that its bird-like features – for instance fused handbones – evolved independently to help it dig. Then in 1996 another related genus came to light: *Patagonykus*, which shared the single-clawed hand but was twice as large. Comparisons with *Alvarezsaurus* showed that theropod also to be a relation. The Alvarezsauridae family named after that dinosaur had already been established, so *Mononykus* and *Patagonykus* were moved into it and joined later by equally bizarre-looking feathered insectivores such as *Shuvuuia* and *Kol*.

70mya

INSECTIVOROUS

**3.5kg
(8lb)**

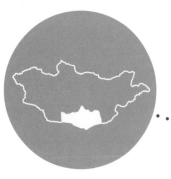

**OMNOGOV,
MONGOLIA**

287

MAASTRICHTIAN	
CAMPANIAN	
SANTONIAN	LATE
CONIACIAN	
TURONIAN	
CENOMANIAN	
ALBIAN	
APTIAN	
BARREMIAN	
HAUTERIVIAN	EARLY
VALANGINIAN	
BERRIASIAN	

75mya

C

CARNIVOROUS

750kg
(1650lb)

UTAH,
USA

288

(teh-RAT-oh-FOE-nee-us)

TERATOPHONEUS CURRIEI

6m
(20ft)

Although 'monstrous murderer' is one of the more frightening dinosaur names coined in recent years, it befitted *Teratophoneus* less well than its descendants; this tyrannosaur was only one-tenth the weight of *Tyrannosaurus*. However, the announcement in 2011 of its discovery in Utah's Grand Staircase-Escalante National Monument, and a year earlier the publication of *Bistahieversor* from New Mexico, helped detail a grey area of the tyrannosaurs' evolution and their distribution in the south-western USA.

Teratophoneus had an unusually short skull containing a reduced number

MAASTRICHTIAN	
CAMPANIAN	LATE
SANTONIAN	
CONIACIAN	
TURONIAN	
CENOMANIAN	
ALBIAN	
APTIAN	
BARREMIAN	EARLY
HAUTERIVIAN	
VALANGINIAN	
BERRIASIAN	

··· **AND** ········· ···· *BISTAHIEVERSOR SEALEYI* ········· ···· ▶

(BIS-tah-HEE-eh-ver-sor)

of teeth, whereas *Bistahieversor* was a bigger beast and actually had more teeth than *Tyrannosaurus*, a startling 64 curved serrated fangs in all. Its name means 'destroyer of Bistahi', which is a Navajo name for the wilderness in New Mexico where several partial skeletons and skulls have been found.

The pair form part of a revival in tyrannosaur discoveries, which began in 2005 over in the eastern USA with the naming of the first one found in North America in 30 years.

75mya

CARNIVOROUS

1000kg
(0.9 tons)

**9m
(30ft)**

**NEW MEXICO,
USA**

····· ▶

AND ·········· ▶

289

CRETACEOUS

MAASTRICHTIAN	
CAMPANIAN	
SANTONIAN	LATE
CONIACIAN	
TURONIAN	
CENOMANIAN	
ALBIAN	
APTIAN	
BARREMIAN	
HAUTERIVIAN	EARLY
VALANGINIAN	
BERRIASIAN	

77mya

C

CARNIVOROUS

900kg
(1980lb)

ALABAMA,
USA

290

(AP-ah-LAY-sha-SORE-us)
APPALACHIASAURUS MONTGOMERIENSIS

Appalachiasaurus was discovered in the modern USA's Appalachian mountain region – Montgomery County, Alabama, to be precise – but Appalachia was also the name of the island continent that lay east of the Western Interior Seaway in the Cretaceous. The specimen found was the most complete theropod known from the eastern United States. The fossil came from a sub-adult so full-grown adults would have been bigger than its 7m (23ft) length. It was part of the albertosaurine sub-family of Tyrannosauridae – other members include *Gorgosaurus* and *Albertosaurus*.

Before *Teratophoneus*, *Bistahieversor* and *Appalachiasaurus'* emergence, tyrannosaurs were only known from about 7m years later, living in the western USA.

7m
(23ft)

CRETACEOUS

MAASTRICHTIAN	
CAMPANIAN	
SANTONIAN	LATE
CONIACIAN	
TURONIAN	
CENOMANIAN	
ALBIAN	
APTIAN	
BARREMIAN	EARLY
HAUTERIVIAN	
VALANGINIAN	
BERRIASIAN	

77–74
mya

CARNIVOROUS

2500kg
(2.4 tons)

ALBERTA,
CANADA AND
MONTANA,
USA

292

(das-PLEE-toh-SORE-us)
DASPLETOSAURUS TOROSUS

AS LONG
AS A CESSNA
LIGHT PLANE

9m
(30ft)

Its teeth were even bigger than those of *Tyrannosaurus* in proportion to skull length, though this was a smaller member of the family and lived 10m years earlier. Like its better known relative, *Daspletosaurus* had a bone-breaking bite – its metre-long (3ft 3in) skull was riddled with openings that allowed for powerful muscle attachments. Its eyes faced forward, the better to track its prey's movement; its sense of smell was refined, its hearing acute; all its senses were heightened to ensure that no potential meal escaped its attention. The build was bulky, the short yet powerful arms bore two fingers: with *Daspletosaurus*, the great tyrannosaurs' template was established.

Along with its distant relation *Gorgosaurus*, it ruled the coastal forests of north-western North America. Charles M. Sternberg made the first find in 1921 in Alberta, Canada. In 2005 a large adult, a sub-adult and a juvenile were discovered in Montana together with five hadrosaurs, whose bones bore *Daspletosaurus* bite-marks. It seems possible the tyrannosaurs were hunting in a pack and died while gorging on their prey. But if they cooperated it may be because they knew this to be the most assured way of making a kill, rather than because they were social animals: their bones are known bearing gouges and scratches from other *Daspletosaurus*' teeth. Some experts think it more likely that they congregated in a free-for-all feeding frenzy, much like Komodo Dragons do today, and that like those modern lizards they may have attacked, killed and even eaten each other in the process.

(SHAN-dung-oh-sore-us)

SHANTUNGOSAURUS GIGANTEUS

· · · · · · · · · · · ·▶

CRETACEOUS

MAASTRICHTIAN	
CAMPANIAN	
SANTONIAN	LATE
CONIACIAN	
TURONIAN	
CENOMANIAN	
ALBIAN	
APTIAN	
BARREMIAN	EARLY
HAUTERIVIAN	
VALANGINIAN	
BERRIASIAN	

This flat-headed hadrosaur may well have been the biggest biped of all time. Numerous skeletons are known, which together provide a fair idea of this huge creature's appearance: it had a 1.5m-long (5ft) duck-billed skull, its jaws were embedded with around 1500 tiny, plant-gnashing teeth, and it weighed as much as two elephants. Such a mass would not be unusual for a sauropod but for any other dinosaur it is remarkable. All the fossils emerged from the same quarry in China's Shandong province, which suggests that *Shantungosaurus* lived in herds, perhaps for protection against predators such as *Tarbosaurus*. Like fellow flat-headed hadrosaurs such as *Edmontosaurus*, it had a concave area surrounding the nostrils that may have been covered by a flap of skin that made a noise when inflated, allowing them to signal warnings to one another. Discovered in 1973 by the Chinese palaeontologist Hu, it may have spent much of its time on all fours but could certainly walk on two legs. This is the position held by an imposing model assembled from casts of the various skeletons that looms over visitors to the Shandong Provincial Museum.

70mya

HERBIVOROUS, HIGH BROWSER

15,000kg (14.7 tons)

16m (52ft)

◀ · ▶

SHANDONG PROVINCE, CHINA

MAASTRICHTIAN	
CAMPANIAN	
SANTONIAN	LATE
CONIACIAN	
TURONIAN	
CENOMANIAN	
ALBIAN	
APTIAN	
BARREMIAN	EARLY
HAUTERIVIAN	
VALANGINIAN	
BERRIASIAN	

70mya

C

CARNIVOROUS

20kg
(44lb)

MAHAJANGA
PROVINCE,
NORTHWESTERN
MADAGASCAR

294

(mah-SHEE-ah-kah-SORE-us)

MASIAKASAURUS KNOPFLERI

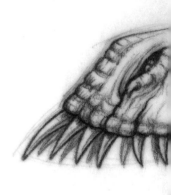

Dagger-like lower front teeth jutting straight out of its mouth mark *Masiakasaurus* out as a little carnivore like no other. It perplexed its finders at first – they were unsure whether it was even a dinosaur until they examined the entire jaw and found that, while the first four teeth gradually inclined from near-horizontal to vertical, the back section was more typical of a theropod. So why did the front teeth develop so strangely? A good rule of thumb in trying to ascertain why extinct animals evolved certain features is to observe the behaviour of any comparable animals alive today. An odd little group of South American marsupials called shrew opossums have similar fangs and use them to prong insects, then using their hind teeth for biting and chewing. It is suspected that the considerably larger *Masiakasaurus* used a similar technique to catch fish and small vertebrates.

It was described in 2001 after being discovered in Madagascar: 'masiaka' is Malagasy for 'vicious', and the species name reflects the fact that the researchers from the University of Utah were listening to Dire Straits as they unearthed the jaws, some limb bones and vertebrae. In 2011 a new find meant that two-thirds of the skeleton is now known, greatly improving experts' understanding. It was a kind of abelisauroid termed a noasaurid, which was a small group forming a twig from the branch leading to the powerful, streamlined abelisaurids such as *Carnotaurus* and *Abelisaurus*. Those advanced genera were notable for their tiny arms and blunt skulls but in the small, more primitive *Masiakasaurus* the arms were designed for grasping, while the skull was low and elongated.

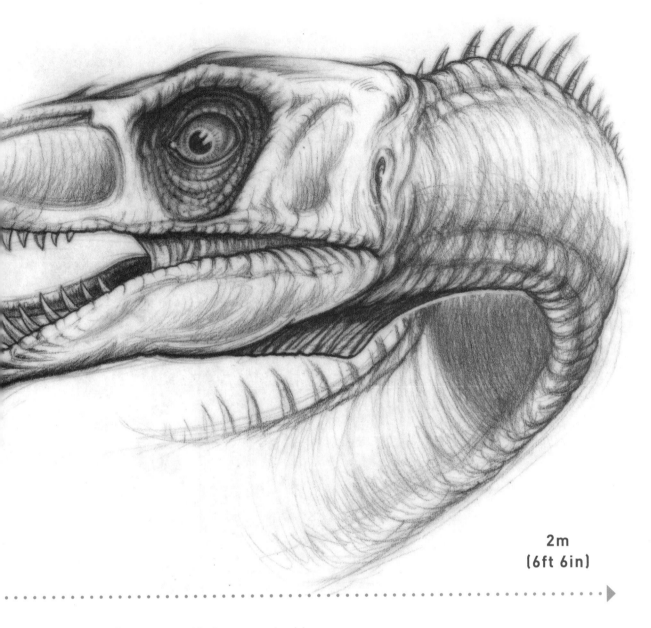

**2m
(6ft 6in)**

By the Late Cretaceous Madagascar had been
an island for 20m years, having split from
India and drifted toward Africa. Then as now it
was rich with strange animal life: *Simosuchus*,
a little pug-nosed plant-eating crocodile,
for instance, or *Rahonavis*, a controversial
creature that may have been a bird or possibly
a flying dromaeosaurid dinosaur. Certainly
among the Madagascan dinosaurs was
Majungasaurus, which probably preyed on
Masiakasaurus judging by the tooth-marks on
one of its fossil bones.

CRETACEOUS

MAASTRICHTIAN	
CAMPANIAN	
SANTONIAN	LATE
CONIACIAN	
TURONIAN	
CENOMANIAN	
ALBIAN	
APTIAN	
BARREMIAN	EARLY
HAUTERIVIAN	
VALANGINIAN	
BERRIASIAN	

70–65.5
mya

C

CARNIVOROUS

750kg
(1650lb)

MADAGASCAR

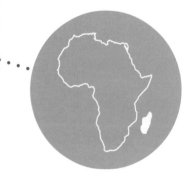

296

(mah-JOONG-ah-SORE-us)

MAJUNGASAURUS CRENATISSIMUS

While *Majungasaurus* may have eaten *Masiakasaurus*, it definitely ate its own kind – for this is the only dinosaur to have been confirmed as a cannibal (though there's some evidence that *Daspletosaurus* (page 292) and *Tyrannosaurus* (page 322) were as well). Numerous *Majungasaurus* bones bear marks matching the size, spacing and serration of its thick, powerful teeth, and no other carnivore its size is known to have lived on Madagascar. Identical chips and gouges are also found on the bones of the island's sauropods; while *Majungasaurus* wasn't huge by theropod standards, it was the island's dominant predator during the Cretaceous.

It had a broad skull for an abelisaurid, a small horn on its head and an especially horizontal posture. In 2007 American palaeontologists Scott Sampson and Lawrence Witmer digitally scanned its skull, and discovered this posture by examining the ear canals, which help with balance. The inner ear contains three of these, of which one, the lateral canal, is parallel with the ground in the head's alert position – think of it like a natural spirit level within an animal's head. When they positioned a *Majungasaurus* skull with the canal parallel to the ground, its skull was almost horizontal; by comparison most theropods' heads inclined downwards when alert. Witmer's 3-D digital scans of dinosaurs' skulls (created via a method properly called X-ray computed tomography) have created a detailed understanding of their brains' shape and size, and thus of their likely posture, sensory perception and behaviour. The method is also far less labour-intensive and destructive than sawing through great chunks of stone!

Like most other abelisaurids, *Majungasaurus'* arms were very small; its strongly muscled jaws were its means of attack. The teeth were perfect for gripping into sauropods' flesh, and the powerful neck helped it shake its victims while it held them, inflicting fatal wounds. It could then rip lumps from the carcass once the kill was complete. Compare this with the slender, knife-like teeth of carcharodontosaurs, better suited to slashing slices of meat from living animals' flanks.

Its stout, muscular legs were unusually short, giving it a low centre of gravity and a slow running speed – but it only needed to be faster than the sauropods that it pursued, such as *Rapetosaurus*.

Majungasaurus' remains were discovered in 1895 by French soldiers who were trying to recover Madagascar from the British, and the following year it was mistakenly assigned to *Megalosaurus*. New partial finds in 1955 led to it being named *Majungasaurus*, after the fossil site's location in Madagascar, but a detailed picture only developed with the discovery of an exquisitely preserved skull in 1996. Since then more specimens have combined to provide knowledge of almost the entire skeleton.

**7m
(23ft)**

MAASTRICHTIAN	
CAMPANIAN	
SANTONIAN	
CONIACIAN	LATE
TURONIAN	
CENOMANIAN	
ALBIAN	
APTIAN	
BARREMIAN	
HAUTERIVIAN	EARLY
VALANGINIAN	
BERRIASIAN	

66mya

HERBIVOROUS

45kg
(99lb)

SOUTH
DAKOTA,
USA

298

(dray-COR-ex)

DRACOREX HOGWARTSIA ···· AND ···

As anyone familiar with the Harry Potter books or movies will know, pupils at the Hogwarts School of Witchcraft and Wizardry often have to contend with the unwelcome attentions of fire-breathing dragons. When Bob Bakker needed a name for a pachycephalosaur in 2006 he was inspired by its resemblance to a dragon, a fact remarked upon by many young visitors who'd seen the skull on display at the Children's Museum of Indianapolis.

The 'dragon king of Hogwarts' was a spiky-faced ornithischian that lived in the damp forests of the USA at the very end of the dinosaurs' era. But while it has a notably unconventional name, of more substantial interest is the question of whether it is really a new discovery – an issue with wider implications for many other dinosaurs. Three years after it was described, the American palaeontologist Jack Horner argued that *Dracorex* wasn't a new dinosaur at all, just a juvenile of the bone-headed *Pachycephalosaurus* whose skull hadn't completed growing.

3m
(10ft)

(stij-ee-MOW-lock)

STYGIMOLOCH SPINIFER

**3m
(10ft)**

CRETACEOUS

MAASTRICHTIAN	
CAMPANIAN	
SANTONIAN	LATE
CONIACIAN	
TURONIAN	
CENOMANIAN	
ALBIAN	
APTIAN	
BARREMIAN	EARLY
HAUTERIVIAN	
VALANGINIAN	
BERRIASIAN	

67–65 mya

HERBIVOROUS

300kg
(700lb)

**MONTANA,
WYOMING, AND
SOUTH AND
NORTH DAKOTA,
USA**

The same went for *Stygimoloch*, another little horned dragon-like biped that browsed on bushes and ferns around Late Cretaceous South Dakota. Horner's findings were supported by a team of researchers from University College Berkeley, the University of California, and his own academic base, the University of Montana.

Horner cited the 1970s studies by Peter Dodson of the University of Pennsylvania, which pointed out that the cassowary bird grows to 80 per cent of its adult size before growing the distinctive large crest on its head. If dinosaurs had the same growth patterns then it made sense to think of them as 'shape-shifters', Horner argued, their facial ornamentation assuming various forms as they developed. In his opinion *Dracorex* is a young juvenile, *Stygimoloch* a near-adult and *Pachycephalosaurus* a grown adult of the same species. *Dracorex* had spikes on its snout and skull, *Stygimoloch* had fewer spikes and the beginnings of a thickened cranium, and *Pachycephalosaurus* had lost all the spikes and developed a great bony bump over its head up to 25cm (10in) thick.

AND

PACHYCEPHALOSAURUS

The three bone-headed dinosaurs *Dracorex*, *Stygimoloch*, and *Pachycephalosaurus* ... or are they all *Pachycephalosaurus* in three growth stages?

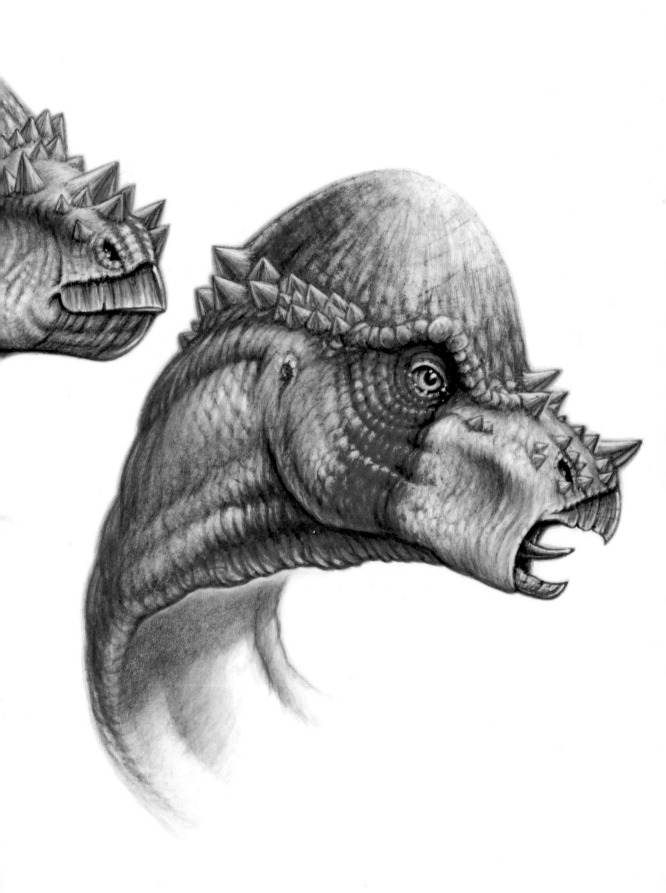

MAASTRICHTIAN	
CAMPANIAN	
SANTONIAN	LATE
CONIACIAN	
TURONIAN	
CENOMANIAN	
ALBIAN	
APTIAN	
BARREMIAN	EARLY
HAUTERIVIAN	
VALANGINIAN	
BERRIASIAN	

68–65
mya

HERBIVOROUS

450kg
(990lb)

MONTANA,
WYOMING,
SOUTH DAKOTA,
USA

302

(PACK-ee-KEFF-ah-low-SORE-us)

PACHYCEPHALOSAURUS WYOMINGENSIS

4.5m
(15ft)

Horner and his colleagues' headline-grabbing conclusion was that perhaps one-third of all known species should be dismissed as juvenile forms of other dinosaurs. Other examples they cited include *Torosaurus*, which he says is a fully grown *Triceratops* (page 316), and *Anatotitan*, supposedly a larger *Edmontosaurus*. This has all proved highly controversial and the palaeontologists who named the genera that Horner wants to see expunged remain adamant that their work is valid. The debate is likely to run and run.

(AL-a-mo-SORE-us)

ALAMOSAURUS SANJUANENSIS

MAASTRICHTIAN	
CAMPANIAN	
SANTONIAN	LATE
CONIACIAN	
TURONIAN	
CENOMANIAN	
ALBIAN	
APTIAN	
BARREMIAN	EARLY
HAUTERIVIAN	
VALANGINIAN	
BERRIASIAN	

This titanosaur is the largest dinosaur known from North America, though at around 20m (65ft) long and 8.5m (28ft) tall it was no giant among sauropods. In 2011 came a contentious find that was claimed to make it drastically larger: according to University of Pennsylvania researchers the two massive vertebrae and a thighbone found in New Mexico suggest a dinosaur comparable with *Argentinosaurus* – but other experts are sceptical about the calculations. An earlier discovery in Texas of an adult and two juvenile fossils suggests that it may have lived in family groups, and a study suggests that within Texas at any one time the *Alamosaurus* population probably numbered around 350,000. Texas is such a huge area that that is only one for

every 0.7 sq miles ... but still, this is a lot of big dinosaurs. The name, incidentally, does not come from the Battle of the Alamo, the pivotal moment in the state's history in which Texans and Mexicans fought over ownership of the site of a Roman Catholic mission, but from the Ojo Alamo rock formation in which the fossils were found. Charles Gilmore described and named the dinosaur in 1922.

70–65 mya

HERBIVOROUS

16,000kg (15.7 tons)

ACROSS SOUTH-WESTERN USA

20m (66ft)

MAASTRICHTIAN	
CAMPANIAN	
SANTONIAN	LATE
CONIACIAN	
TURONIAN	
CENOMANIAN	
ALBIAN	
APTIAN	
BARREMIAN	EARLY
HAUTERIVIAN	
VALANGINIAN	
BERRIASIAN	

70–65 mya

H

HERBIVOROUS

15,000kg
(14.7 tons)

INDIA

(ISS-ee-SORE-us)

ISISAURUS COLBERTI

18m (59ft)

Surely one of the weirdest-looking sauropods, this hefty titanosaur had a relatively short but strangely deep and slender neck. The purpose of this modification isn't certain, though one suggestion is that it would have borne markings or colours for display purposes. Discovered in central India (it was named in honour of the Indian Statistical Institute, which holds numerous fossils), *Isisaurus* was originally described as a species of *Titanosaurus* but became a new genus in 2003. Two years later scientists analysed coprolites believed to have come from *Isisaurus* and found that it had a diverse diet, eating leaves from various kinds of tree. Its unusually long front legs would have helped it feed from the topmost branches. It lived alongside fellow titanosaur *Jainosaurus*, and both would have been hunted by *Indosuchus* and *Rajasaurus* (page 306).

DINO-DUNG = GOLD DUST

Just as their bones turned into stone over the aeons, so too did some of their droppings. Fossilised animal faeces are called coprolites and they are actually of great scientific value. Naturally enough these gigantic beasts produced faeces to match: the largest ever found is 17in long by 6in across. Palaeontologists suspect that particular coprolite to have been passed by a *T. rex* around 65mya. Their belief that it came from a *T. rex* derives from the presence of chips of bone – in fact certain palaeontologists spend the majority of their time examining excrement for an insight into the diet of the creature that produced it. Coprolites can reveal whether a dinosaur was carnivorous, omnivorous or herbivorous – and where plant matter is found, studying dino-dung provides an idea of the kind of foliage that grew in that dinosaur's habitat. Sometimes this can reveal the presence of a certain plant millions of years earlier than it had previously been known, changing scientists' understanding of plant evolution. While it is usually hard to determine which species of dinosaur left a coprolite, sometimes they are actually preserved within the intestinal area of a fossilised skeleton. Mary Anning found some in the ichthyosaur remains that she discovered in Dorset in the early 19th century, immediately providing an understanding of those marine reptiles' diet.

In fact, for palaeontologists, coprolites are like gold dust – they are extremely hard to find and more revealing than fossilised bones in terms of illustrating dinosaurs' eating habits.

MAASTRICHTIAN	
CAMPANIAN	
SANTONIAN	LATE
CONIACIAN	
TURONIAN	
CENOMANIAN	
ALBIAN	
APTIAN	
BARREMIAN	EARLY
HAUTERIVIAN	
VALANGINIAN	
BERRIASIAN	

67mya

CARNIVOROUS

4000kg
(3.9 tons)

INDIA

(RAAJ-a-SORE-us)

RAJASAURUS NARMADENSIS

With a 'crown' of small horns around the top of its skull, this muscular predator struck its discoverers as having a regal air – hence a name meaning 'Indian king lizard'. The horns were probably for display and for headbutting rival *Rajasaurus*, a common behaviour among the bony-headed abelisaurids.

In 1981, workers at a cement quarry in Gujarat grew intrigued by a collection of limestone 'balls' emerging from the rockface. These turned out to be dinosaur eggs, and over the next three years the Geological Survey of India drew out a complete skull and partial skeleton from a sandstone bed lying beneath. In 2003 an Indian and American team eventually revealed them to represent a

new theropod. The skull's fine preservation owes much to the widespread volcanic activity in India's Deccan steppes at this time (often considered a contributory factor in the non-avian dinosaurs' extinction – see page 336). While this *Rajasaurus* would have spent its life in a land of rivers, lakes and greenery, after it died this landscape was obliterated by an orange-glowing flood of molten rock. The sediments enclosing its bones were covered by the igneous rock formed when the lava flow cooled, which served to preserve *Rajasaurus* beautifully until its emergence 67m years later, at which point it sparked huge interest in India and beyond.

11m
(36ft)

FRANZ, BARON NOPCSA
Hungarian (1877–1933)

One of the most extraordinary figures
of his time, Nopcsa was a Transylvanian
aristocrat, an adventurer, a spy, a would-
be king of Albania and finally a murderer
– but also a serious palaeontologist
at a time when few others considered
the discipline worth pursuing. He was
a brilliant scholar, giving lectures on
pterosaur physiognomy by the age of 22.
One of his most enduring contributions
to palaeontology was the now-vindicated
theory that Romanian dinosaurs had
evolved as dwarf forms because they
lived on what was then an island (see
page 308). Nopcsa also advocated a link
between birds and dinosaurs, and tried
to understand how dinosaurs' organs and
soft tissues worked rather than simply
focusing on skeletal structure. He became
an advocate of Albanian nationalism –
at that time the region was subsumed
into the Ottoman Empire – and tried to
convince the Albanians to install him as
their monarch, but failed. He fell into
debt, sold his fossil collection to London's
Natural History Museum, and later killed
his Albanian boyfriend at their apartment
in Vienna before shooting himself.

DWARF DINOSAURS

Imagine an island of dwarf dinosaurs, populated by sauropods and ornithopods that look familiar but are modelled in miniature.

Towards the end of the Cretaceous period much of Europe was below sea level and an archipelago of islands lay scattered across eastern Europe and the Mediterranean coast. One is referred to today as Hateg Island, a 30,000-square-mile verdant land rich in plant and animal life. The fossil record reveals insects, fish, frogs, lizards, birds, mammals – and dinosaurs. Around a century ago Baron Franz Nopcsa began finding their fossils on his land in Romania and noticing their relation to dinosaurs found in older rocks in England, Germany and North America, but with a strange difference: the Hateg dinosaurs were all far smaller. Three genera, the sauropod *Magyarosaurus* and the ornithopods *Telmatosaurus* and *Zalmoxes*, were half the length of their nearest relatives elsewhere.

He proposed the notion that they were dwarfs rather than juveniles, and in 2010 a team led by Bristol University's Professor Mike Benton finally put the theory to the test. A close assessment of the bones revealed that they were fused and thus unable to grow further, confirming that they did indeed belong to adults.

It fits in with a debate within evolutionary ecology about the 'island rule', which states that large animals isolated on islands tend to evolve to become smaller as they cope with competition for a restricted food supply. Dwarf elephants are known to have lived on Mediterranean islands within the past tens of thousands of years, Professor Benton noted. The isolation can also produce some strange specialisations, such as those seen in *Balaur*.

(bal-AY-ur)

BALAUR BONDOC

MAASTRICHTIAN

CAMPANIAN

SANTONIAN

CONIACIAN

TURONIAN

CENOMANIAN

ALBIAN

APTIAN

BARREMIAN

HAUTERIVIAN

VALANGINIAN

BERRIASIAN

LATE

EARLY

Balaur was one of the most bizarre little theropods yet discovered, and the most complete meat-eater known so far from Europe in the non-avian dinosaurs' final 60m years. It inhabited an island in what's now Romania (see opposite) and was superficially similar to other dromaeosaurs, those vicious, highly evolved feathered hunters with a sickle claw on each foot. But *Balaur* had two sickle claws on each foot, and a claw on each of its hands too, and its limbs were far more muscular than those of *Velociraptor*, its closest known relative. *Balaur* was only small but it was beefed up and had an array of deadly weapons at its disposal. Its hands were relatively weak so it's thought that it used its toe-claws for disembowelling its prey. In the words of Stephen Brusatte, one of the research team who described it, *Balaur* was 'probably more of a kickboxer than a sprinter, and it might have been able to take down larger animals than itself, as many carnivores do today'. Its remains caused much bafflement when they were unearthed in 1997 but, after a thorough investigation, 2010 saw it announced as Europe's most startling dinosaur find for years. The name is that of an evil dragon from Romanian mythology, and 'bondoc' is from the Turkish for 'stocky', which also alludes to the likely Asian origin of this dinosaur's ancestors.

70–65 mya

C

CARNIVOROUS

11kg (25lb)

ROMANIA

2m (7ft)

(illustrated overleaf)

CRETACEOUS

MAASTRICHTIAN	
CAMPANIAN	
SANTONIAN	LATE
CONIACIAN	
TURONIAN	
CENOMANIAN	
ALBIAN	
APTIAN	
BARREMIAN	
HAUTERIVIAN	EARLY
VALANGINIAN	
BERRIASIAN	

72–69
mya

C

CARNIVOROUS

2000kg
(1.9 tons)

ARGENTINA

312

(car-no-TORE-us)
CARNOTAURUS SASTREI

The 'flesh-eating bull' may just have been the fastest of all the terrifying major carnivores that ruled the Late Cretaceous. For while it has traditionally gained attention for the horns on its brow, research by Scott Persons and Phil Currie published in 2011 suggested that *Carnotaurus*' most unusual feature lay at the other end of its body. This was by any measure a pretty strange theropod – it had a notably long neck and forearms so withered that they make *T. rex*'s look well developed. But it turns out its most important innovation was its tail. In a paper that they gave the unusually exciting title 'Dinosaur Speed Demon', Persons and Currie looked at a series of interlocking rib-like bones at the top of its tail and interpreted a series of distinct ridges as 'attachment scars', marking where a pair of muscles called the caudofemoralis joined to the bone. Tendons attached the other end of each muscle to the two thighbones. They created a digital model of the tail muscles and found that they would have been huge, bigger than those of any other theropod. Contracting them served to pull the legs upwards while running – so the tail gave *Carnotaurus*' thighs extraordinary explosive power, allowing it to bound along at speeds of up to 30mph. Because the tail was so reinforced with bone, it was unusually stiff, and this meant

that *Carnotaurus* had trouble changing direction. That was the only consolation for the smaller herbivores in its territory.

'The tail was rigid, making it difficult for the hunter to make quick, fluid turns,' Persons stated when publishing the paper. 'Imagine yourself as a small plant-eating dinosaur on the floodplains of prehistoric Argentina, and you are unlucky enough to find yourself being charged by a hungry *Carnotaurus*. Your best bet is to make a lot of quick turns, because you couldn't beat *Carnotaurus* in a straight sprint.'

This was the latest and most specialised of the abelisaurs, a group of short-legged and relatively streamlined theropods that dominated the southern hemisphere alongside the huge carcharodontosaurs, whereas the bulky tyrannosaurs held control in the north. Abelisaurs were most distinctive for the ornamentation on their skulls (others include *Rugops*, page 229, and *Rajasaurus*, page 306). There is no consensus on the purpose of *Carnotaurus*' horns; some experts suggest they were for butting rivals of the same species, others say its skull was too weak to withstand such an impact, and others suspect they were for display or even attacking prey.

One certainty is that we can discount the idea of this large

7.5m (25ft)

theropod being feathered. A *Carnotaurus* fossil found in Argentina provides the most complete skin impression we have of any carnivorous dinosaur, revealing it to be covered by rounded scales with rows of larger scutes running along its flanks. Its forward-facing eyes probably gave it binocular vision – useful for a hunter tracking its prey – but some mystery surrounds what it preyed upon and how. Its skull was short and its lower jaw lightweight, suggesting it could bite quickly but with less power than the biggest theropods. It may have used its speed to pursue bird-like plant-eaters, leaving the big sauropods to the likes of *Giganotosaurus*. Those massive carcharodontosaurs were presumably not fussy eaters, so it is plausible that they sometimes turned their sights on smaller carnivores. Evading their attentions would have been another good use of *Carnotaurus'* sprinting ability.

CRETACEOUS

MAASTRICHTIAN	
CAMPANIAN	
SANTONIAN	LATE
CONIACIAN	
TURONIAN	
CENOMANIAN	
ALBIAN	
APTIAN	
BARREMIAN	EARLY
HAUTERIVIAN	
VALANGINIAN	
BERRIASIAN	

70–68
mya

HERBIVOROUS

4.5kg
(10lb)

**SHANDUNG
PROVINCE,
CHINA**

(MY-crow-PACK-ee-KEFF-a-lo-SORE-us)

MICROPACHYCEPHALOSAURUS HONGTUYANENSIS

The longest name given to any dinosaur was granted in 1978 to one of the smallest kinds known, a bipedal herbivore the size of a large rabbit that lived in Late Cretaceous China. Its name is simple enough when you deconstruct it: micro-pachy-cephalo-saurus means 'small-thick-headed-lizard'. From this you would assume that it was a little bone-headed dinosaur, a diminutive version of the better known *Pachycephalosaurus*. That is what the Chinese palaeontologist Dong Zhiming thought when he named it. But its remains are very fragmentary, just a pelvis and a partial skull that doesn't retain the cranium, which would determine whether it was a bone-head or not. Research in 2011 led by Richard Butler, of Cambridge University and the Natural History Museum, concluded it was actually a primitive member of the ceratopsians.

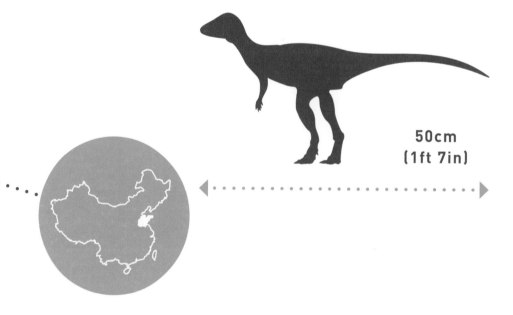

50cm
(1ft 7in)

(an-KILE-oh-SORE-us)

ANKYLOSAURUS MAGNIVENTRIS

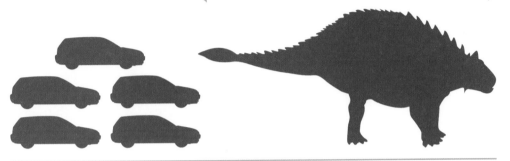

7m
(23ft)

MAASTRICHTIAN		
CAMPANIAN		
SANTONIAN	LATE	
CONIACIAN		
TURONIAN		
CENOMANIAN		
ALBIAN		
APTIAN		
BARREMIAN		
HAUTERIVIAN	EARLY	
VALANGINIAN		
BERRIASIAN		

68–65
mya

HERBIVOROUS

3000kg
(2.9 tons)

**MONTANA
AND WYOMING,
USA, AND
ALBERTA,
CANADA**

Picture a broad, squat beast the length and height of a minibus, covered with knobbly armour such as you'd find on a crocodile's back, and wielding a tail ending with a club that could smash through bone... you have something resembling *Ankylosaurus*. These creatures were tough, which they needed to be – they had predators such as *Tyrannosaurus* trying to kill and eat them. *Ankylosaurus* lived in what's now western North America in the Late Cretaceous, but it gives its name to the ankylosaur group, which was far more widespread and long-lived. Ankylosaur fossils have been found on every continent but Africa and date back to the Early Jurassic. The fact that *Ankylosaurus* and its relatives survived to the end of the Cretaceous is proof of their hardiness. By the later stages of their evolution one variety of ankylosaur had even developed armoured eyelids. *Ankylosaurus* remains tend to be found alone, suggesting they were not herding animals. They grazed slowly on vegetation and were not overly blessed with brainpower: for this reason they are sometime dubbed the cows of the Cretaceous.

MAASTRICHTIAN
CAMPANIAN
SANTONIAN
CONIACIAN
TURONIAN
CENOMANIAN
ALBIAN
APTIAN
BARREMIAN
HAUTERIVIAN
VALANGINIAN
BERRIASIAN

LATE

EARLY

68–65
mya

H

HERBIVOROUS

11,000kg
(10.8 tons)

WESTERN
USA

316

(try-SERR-ah-tops)

TRICERATOPS HORRIDUS

With two great horns sprouting from its brow, a smaller one atop its snout and an immense rounded frill masking the back of its neck, this is one of the true icons of the Mesozoic era. Anyone with a passing interest in dinosaurs knows *Triceratops*, along with a few others: *Tyrannosaurus*, *Stegosaurus* and *Diplodocus*, for instance, each of which is easily pictured. But while the familiar image of *Triceratops* is of a huge herbivore, its mighty skull held high above an adult human's head, it now seems that the fully grown animal was larger still. Othniel Marsh named *T. horridus* in 1889 and *T. prorsus* in 1890, but the following year he described a ceratopsid with three differently angled horns placed upon one of the longest known skulls of any land animal, reaching 2.6m (8ft 6in) from the tip of its beaked mouth to the end of its frill. He called it *Torosaurus*. In 2010 the American palaeontologists Jack Horner and John Scannella concluded that *Torosaurus* is *Triceratops* in its fully grown form; that is to say that the famous *Triceratops* is merely the juvenile and 'sub-adult' form of this even more spectacular animal.

Torosaurus is notable for the large holes in its neck-shield, and Scannella's closer examination

of *Triceratops*' seemingly solid shield revealed the bone thinning where these gaps would form. If this is the case, it strengthens the notion that *Triceratops*' frill had little defensive use but was instead for sexual display, the most extravagant examples denoting the alpha males, just like stags with their antlers. Horner and Scannella's conclusions are not universally accepted but, if they are right, the famous *Triceratops* name would still stay with us: its name takes priority as Marsh named it two years before *Torosaurus*.

Whatever the name we apply to it today, this animal was one of the most abundant herbivores of the Late Cretaceous. The western USA's Hell Creek Formation rocks are forever yielding new specimens; 47 *Triceratops* skulls emerged in the first decade of this century alone. Unlike some other large herbivores there is no evidence that *Triceratops* herded; the hundreds of specimens known have all been found in isolation with the exception of three juveniles. Its predators are known to have included *Tyrannosaurus*, as the carnivore's tooth-marks have been found on its skeletons. Fossilised skin impressions are known featuring small round

9m
(30ft)

holes, which some experts suspect may have sprouted bristles, as in fellow ornithischians *Tianyulong* and *Psittacosaurus* (page 183). *Triceratops* is thought to have endured until the great extinction that marked the closure of the dinosaurs' era.

Other species: *T. prorsus*

CRETACEOUS

MAASTRICHTIAN	
CAMPANIAN	
SANTONIAN	LATE
CONIACIAN	
TURONIAN	
CENOMANIAN	
ALBIAN	
APTIAN	
BARREMIAN	EARLY
HAUTERIVIAN	
VALANGINIAN	
BERRIASIAN	

73–70
mya

H

HERBIVOROUS

4000kg
(3.9 tons)

NORTH SLOPE,
ALASKA

(pack-ee-RHINE-oh-SORE-us)

PACHYRHINOSAURUS PEROTORUM

Its appearance was unexceptional – somewhat like a *Triceratops* with a rugged bump instead of a nasal horn – but its discovery in Alaska in 2006 was important. During the Late Cretaceous, the uppermost fringes of North America were even further north than today, meaning that *Pachyrhinosaurus* lived through a winter of six months' darkness and freezing temperatures. The discovery of its skull high on a hillside above the Colville River emphasised the diversity of dinosaurs that managed to live in polar conditions. This ceratopsian would have trudged through the wintry forests cropping plant matter and seeking to evade the attentions of carnivores such as the tyrannosaurid *Gorgosaurus* and the deinonychosaurs *Dromaeosaurus* and *Troodon*, all of whose fossils have emerged from the same Prince Creek Formation in North Slope, Alaska.

This was the third *Pachyrhinosaurus* species discovered – the others were found in Canada – and its specific name honours the former US presidential candidate H Ross Perot. His children donated $50m to the Dallas Museum of Nature and Science, whose researchers found and studied the fossil before publishing its description in 2011.

8m
(26ft)

(AL-ee-oh-RAY-mus)

ALIORAMUS REMOTUS

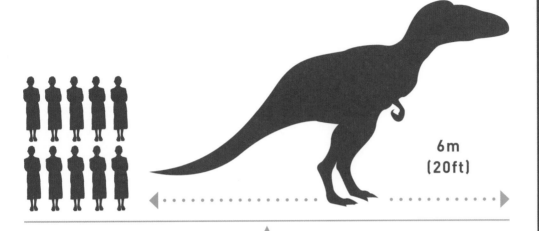

6m
(20ft)

CRETACEOUS

MAASTRICHTIAN	
CAMPANIAN	
SANTONIAN	LATE
CONIACIAN	
TURONIAN	
CENOMANIAN	
ALBIAN	
APTIAN	
BARREMIAN	EARLY
HAUTERIVIAN	
VALANGINIAN	
BERRIASIAN	

70–65
mya

CARNIVOROUS

680kg
(1500lb)

BAYANKHONGOR,
MONGOLIA

Its elongated head marks this as an unusual tyrannosaurid given that better known genera had deep and blunt skulls. *Alioramus*' snout bore five ridges, each a centimetre high, and its narrow jaws contained either 76 or 78 teeth, which either way is more than any other member of the family. Russian palaeontologist Sergei Kurzanov discovered the first species in Mongolia in the early 1970s, later granting a name meaning 'removed other branch' in reference to its differences from American tyrannosaurids. Kurzanov's specimen consisted of a skull and two handbones, but in 2009 the American Stephen Brusatte and colleagues named a new species, *A. altai*, from a more complete skeleton. This confirmed *Alioramus* to be a two-fingered bipedal predator around the height of a human, probably closely related to the larger Asian tyrannosaurid *Tarbosaurus*. Both species were named from juvenile specimens so until an adult *Alioramus* fossil is found, its fully grown size can only be estimated.

319

MAASTRICHTIAN	
CAMPANIAN	
SANTONIAN	
CONIACIAN	LATE
TURONIAN	
CENOMANIAN	
ALBIAN	
APTIAN	
BARREMIAN	
HAUTERIVIAN	EARLY
VALANGINIAN	
BERRIASIAN	

70mya

H

HERBIVOROUS

3000kg
(2.9 tons)

TSINGTAO,
SHANDONG
PROVINCE,
CHINA

320

(ching-dow-SORE-us)

TSINTAOSAURUS SPINORHINUS

The 'unicorn dinosaur' gets its nickname for an obvious reason. It was an elephant-sized member of the hadrosaurs, the family of duck-billed herbivores known for sporting a variety of elaborate skull ornamentation: think of *Parasaurolophus* with its long tubular crest protruding backwards, containing pipes through which it could create a resonant call. But *Tsintaosaurus'* crest speared forwards from its brow and was solid so couldn't have shared this purpose. Proposed functions include helping regulate its temperature, marking adults as sexually mature or helping to heighten its sense of smell. The protuberance's true appearance is hard to gauge, as in life it was probably sheathed in keratin and perhaps supported a sail of skin running down to the neck – in which case it would have looked far less like a unicorn's, but still very strange.

C C Young named *Tsintaosaurus* in 1958. Like other hadrosaurs it probably lived in herds and was primarily quadrupedal but able to run on its hind legs when necessary. Its battery of self-sharpening teeth suited a diet of tough vegetation such as cycads and conifers.

**10m
(33ft)**

(GIL-more-oh-SORE-us)

GILMOREOSAURUS MONGOLIENSIS

6m
(20ft)

Gilmoreosaurus' most interesting feature for palaeontologists is that, along with certain other hadrosaurs, it is one of the few dinosaurs known to have suffered from cancer: its fossilised vertebrae show signs of having harboured tumours. This powerfully built biped, a primitive Asian branch of the duck-billed dinosaurs, roamed through woodlands munching on vegetation. The American palaeontologist George Olsen unearthed the first *Gilmoreosaurus* fossils during an expedition to China and Mongolia in 1923. The bones were scattered around, making clear identification difficult, and were misidentified as belonging to *Mandschurosaurus*, another hadrosaur. In 1979 it was deemed sufficiently distinct to be given its own genus, named in honour of the American palaeontologist Charles W Gilmore (1874–1945).

CRETACEOUS

MAASTRICHTIAN	
CAMPANIAN	
SANTONIAN	LATE
CONIACIAN	
TURONIAN	
CENOMANIAN	
ALBIAN	
APTIAN	
BARREMIAN	
HAUTERIVIAN	EARLY
VALANGINIAN	
BERRIASIAN	

76–70
mya

HERBIVOROUS

1200kg
(1.1 tons)

INNER
MONGOLIA,
NORTHERN
CHINA

321

MAASTRICHTIAN	
CAMPANIAN	
SANTONIAN	LATE
CONIACIAN	
TURONIAN	
CENOMANIAN	
ALBIAN	
APTIAN	
BARREMIAN	EARLY
HAUTERIVIAN	
VALANGINIAN	
BERRIASIAN	

68–65 mya

CARNIVOROUS,
HUNTING AND
SCAVENGING

6000kg
(5.9 tons)

WESTERN
NORTH
AMERICA

322

(tie-RAN-oh-SORE-us)

TYRANNOSAURUS REX

Since Barnum Brown teased its bones from the badlands of Montana in 1902 other theropods have emerged that were bigger and faster, but still this horrific killer stands alone: the most famous dinosaur by far, and the only one best known by its full title. No one speaks of *Triceratops horridus* or *Diplodocus longus*, but the name coined by Henry Fairfield Osborn in 1905 combines an irresistible rhythm with a hint of its menace: *Tyrannosaurus rex*, the tyrant king. Its discovery soon made

headlines: 'The absolute warlord of the Earth' was one description in the *New York Times*, along with 'a royal man-eater of the jungle' – an early example of the irresistible urge to imagine the terror *Tyrannosaurus* could unleash on humans, despite the 63m-year gap between its demise and the moment when our ancestors lifted their knuckles from the ground and began walking on two legs. The *Tyrannosaurus* that shot to fame was an upright beast that charged alone through forests and

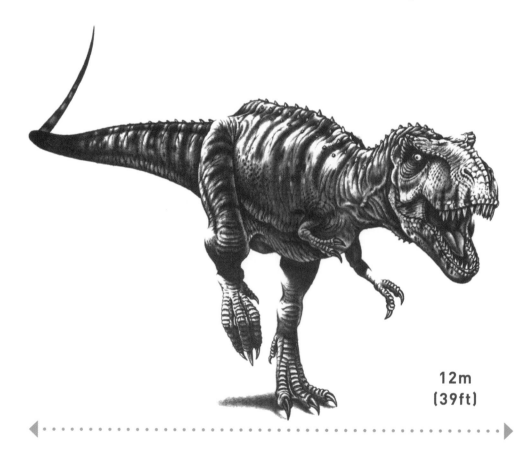

12m
(39ft)

across plains lunging at whichever unfortunate creatures it decided to make its prey, snapping their bones with a sickening crunch of its 1.5m (5ft) jaws, then consuming their carcasses while blood dripped down its scaly jowls. In recent years however the picture has changed. The creature remains one of the most horrific carnivores ever to have lived – those jaws were packed with 60 serrated, banana-sized teeth and it bit with a 1500kg (1.5 ton) force, akin to the weight of a small truck bearing down on each tooth. But we now see its posture as dynamic and horizontal rather than kangaroo-like, its scales may well have been combined with feathers, and there are hints that it hunted in terrifying packs rather than as a solitary killer.

Since *Tyrannosaurus*' discovery around 30 more specimens have been unearthed. The most famous stands 12.8m (42ft) long and 4m (13ft) tall at its hips, and was found in 1990 by fossil-hunter Sue Hendrickson, after whom it is nicknamed 'Sue'. In 1997 it sold for $8.3m at auction, making it the most expensive fossil ever, and today it stands in the Field Museum of Natural History in Chicago.

Other good remains have emerged from Canada all the way down to New Mexico, which combine to create a detailed picture of *Tyrannosaurus*' development, appearance and lifestyle. The differences between adults and juveniles were substantial. By the time it was fully grown – that is, in its early 20s – a *Tyrannosaurus* had undergone a drastic transformation from its adolescent self. Having so many *Tyrannosaurus* fossils at varying stages of maturity, from babies through to the aforementioned 'Sue', means that its growth stages are well documented – and between the ages of 13 and 17, *Tyrannosaurus* underwent an incredible growth spurt in which it put on 1500lb per year. As a juvenile it had blade-like teeth but as it matured they became conical. Its skull grew and thickened until it alone weighed half a ton, and its body bulked out until the lithe, sprinting adolescent had mutated into a massively muscled lumbering beast with an immense appetite for other animals' flesh.

Adults were so 'robust', to use the palaeontological term, that they could probably only pound along at 15mph to at most 25mph, according to the latest calculations. (A pursuing adult tyrannosaur running at the faster end of that scale would still catch most humans, though our top sprinters would probably be safe – when Usain Bolt set the 100m world record at 9.58 seconds, his top speed was an astounding 27.79mph.)

But juveniles were probably faster as they had longer shinbones compared with their thighbones, and the proportionately longer the shinbone is, the better suited an animal is to sprinting. If so, this may tally with the notion that *Tyrannosaurus* hunted in packs. Canadian palaeontologist Philip Currie advocates this theory, although many of his peers feel that more proof is needed. Fellow

CRETACEOUS

MAASTRICHTIAN	
CAMPANIAN	
SANTONIAN	LATE
CONIACIAN	
TURONIAN	
CENOMANIAN	
ALBIAN	
APTIAN	
BARREMIAN	EARLY
HAUTERIVIAN	
VALANGINIAN	
BERRIASIAN	

70–68.5 mya

C

CARNIVOROUS

4000kg (3.9 tons)

MONGOLIA AND NORTHERN CHINA

(TARB-oh-SORE-us)

TARBOSAURUS BATAAR

tyrannosaur *Daspletosaurus* may have done so (see page 292), and the discovery of a group of 68 *Tarbosaurus* skeletons scattered in a remote patch of the Gobi Desert provides, for Currie at least, evidence that tyrannosaurs lived as cooperative social animals, most likely hunting together. For other experts, however, it only confirms that they died together, which could have happened in a natural disaster – perhaps a flood that swept dozens of solitary dinosaurs to this location.

If they did hunt collectively, it is mooted that the speedy youngsters chased the prey down and the massive adults brought up the rear, delivering a fatal bite with their far heavier and more powerful jaws.

There is no doubt that this animal and *Tarbosaurus* (the 'dreadful lizard'), a very similar Asian variety sometimes considered a species of *Tyrannosaurus*, evolved to consume vast quantities of meat, with their tastes spanning from ceratopsians to hadrosaurs.

However, a long-running controversy has surrounded whether *Tyrannosaurus* actually hunted or merely scavenged the carcasses of other dinosaurs' meals. The answer is almost certainly that it did both. Its skull's brain cavity has large olfactory bulbs, which suggest a sense of smell able to detect from afar the scent of carrion abandoned by predatory dinosaurs, but it was quite capable of catching its own meals too. Although an adult *Tyrannosaurus* wasn't a particularly fast runner, it was quick enough to catch most of its potential victims in the Late Cretaceous, and that was all that mattered.

Its large eyes faced forward more than in most dinosaurs, granting an overlap between its fields of vision, and this suggests predatory behaviour because it enabled *Tyrannosaurus* to judge distance and movement, helping pinpoint its prey. Scans of its skull reveal an inner ear that granted

9.5m (31ft)

very fine hearing and balance, and skull cavities that enabled it to hear very low frequency sounds – such as the soft rumble of a herd of *Triceratops* roaming their way across a distant plain.

The existence of *Triceratops* and hadrosaur fossils with healed-over *Tyrannosaurus* bite-marks proves that it attacked living animals, which in these specimens' cases evidently escaped and survived. Few were so lucky. Work in 2012 by palaeontologist Denver Fowler and colleagues at the Museum of the Rockies in Bozeman, Montana, gave a graphic insight into *Tyrannosaurus*' technique for eating ceratopsians. The team examined a collection of *Triceratops* skulls and found that many bore marks on the bony, keratinous frill that suggest biting and pulling after the animal's death. There would have been little flesh to eat there, so they surmised that *Tyrannosaurus* pinioned the corpse with one hefty foot and used its teeth to grip the frill and rip away *Triceratops*' head. After decapitating its victim, it could feast on the nutrient-rich neck muscles that previously lay inaccessible beneath the frill.

As for the possibility that it bore feathers, while some small, circular scale impressions are known, these were only from the underside of a tail. Chinese fossils prove that earlier coelurosaur theropods from which it descended were feathered or fuzzy, so it seems plausible that *Tyrannosaurus* was too. It may be that juvenile tyrannosaurs bore proto-feathers for insulation, as small animals find it harder to

keep warm than larger ones: baby elephants, for instance, are hairier than adults.

But even the mature tyrannosaurs may have retained some feathers for display purposes. The American palaeontologist Thomas Holtz notes that while *Tyrannosaurus* had relatively tiny arms (although well muscled and at 1m (3ft) long, still larger than most humans'), modern flightless birds use their wings for signalling to one another. Perhaps adult tyrannosaurs' arms were feathered and used for show, he says, giving them more use than their skeletons suggest.

Given their arms' size and the fact they bore only two fingers, it's thought that *Tyrannosaurus* attacked with its mouth – and then the hands held its prey in place while the powerful jaws ripped at its flesh.

But while it is famed as the arch devourer of other dinosaurs, being a *Tyrannosaurus* was a dangerous occupation. Although they could live to be 30 years old, it appears from the fossils known that only 2 per cent of them reached the end of their natural lifespan. So this animal, whose life was devoted to killing others, far more often than not itself died a premature death, whether inflicted by fellow tyrannosaurs or the lethal horns of the ceratopsians that it attacked, or the regular flash-flooding that immersed North America's coastal plains beneath water. If any dinosaur perfectly embodied every aspect of the manifold dangers of the Mesozoic world, it was *Tyrannosaurus rex*.

CRETACEOUS

MAASTRICHTIAN	
CAMPANIAN	
SANTONIAN	LATE
CONIACIAN	
TURONIAN	
CENOMANIAN	
ALBIAN	
APTIAN	
BARREMIAN	
HAUTERIVIAN	EARLY
VALANGINIAN	
BERRIASIAN	

MORE AMAZING ANIMALS OF THE CRETACEOUS

(SARK-oh-SOOK-us)

SARCOSUCHUS IMPERATOR

112mya

C

CARNIVOROUS

10,000kg
(9.8 tons)

NIGER,
AFRICA

Dinosaurs passing by the riverside might have spotted the very top of its head jutting through the surface, which allowed *Sarcosuchus* to see and breathe – but the rest of its immense body lay obscured underwater until the moment of attack. Then this gargantuan killer lunged out and clamped its prey in its 1.5m-long (5ft) jaws, in all likelihood employing the modern crocodilians' method of shaking it wildly and pulling it into the water to drown. *Sarcosuchus* was ten times the weight of any crocodile alive today and had a bite-force of eight tonnes: that is, once its jaws closed, trying to push them apart again would be like trying to pick up a particularly large killer whale. These jaws enclosed 132 thick strong teeth that could crunch through bone, and it was the discovery in Niger in 1964 of such teeth, along with some vertebrae and 30cm-long (1ft) armoured scutes, that provided the first intimation of this truly enormous crocodile-like creature from the Cretaceous. The American palaeontologist Paul Sereno led a team back to that inhospitable patch of the Tenere Desert in

1997 and 2000 and found sufficient remains –
several huge skulls, limb bones, more scutes
and vertebrae – to deduce the creature's true
size at last. It is sometimes informally dubbed
'SuperCroc', though *Sarcosuchus* and its fellow
pholidosaurs were not true crocodilians, and
only very distant relations of today's crocodiles.
Seventy scutes set into two long rows made
its back near-impenetrable, rendering
Sarchosuchus an invulnerable killer that could
survive for up to 60 years, throughout which
time it never stopped growing. Even today,
112m years after it fell extinct, it can still
provoke a shudder.

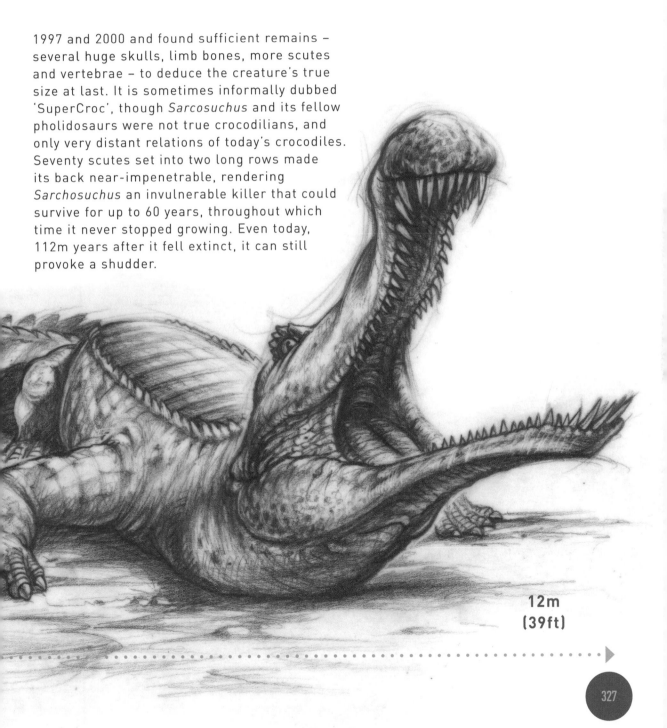

**12m
(39ft)**

MAASTRICHTIAN	
CAMPANIAN	
SANTONIAN	LATE
CONIACIAN	
TURONIAN	
CENOMANIAN	
ALBIAN	
APTIAN	
BARREMIAN	EARLY
HAUTERIVIAN	
VALANGINIAN	
BERRIASIAN	

70–65 mya

C

CARNIVOROUS

uncertain

TEXAS, USA

(KWET-sal-co-AT-lus)

QUETZALCOATLUS NORTHROPI

With a wingspan the length of a bus and a huge, toothless beak, this is the biggest known creature that ever (probably) took to the skies. Known from partial remains found in Texas and named in 1975, *Quetzalcoatlus* was a gigantic pterosaur that, alongside a Romanian specimen referred to as *Hatzegopteryx*, represents a type known as the azhdarchids. These immense creatures' weight is a contested matter, with most estimates varying from 250kg (550lb) down to as little as 70kg (150lb).

The latter is similar to an average human adult but in an animal which, resting on its four limbs, would stand almost eye-to-eye with a giraffe.

The question of its weight has led some to doubt whether it could actually fly. But in recent years researchers have re-examined its arm bones and concluded they were so thickened that they could have been strong enough to propel even a 250kg (550lb) animal into the air. It would have done so from a crouching, four-legged take-off position, according to computer models that show that adopting such a posture would have created enough explosive energy in its limbs to get it airborne. However, a conclusive answer will only emerge with the discovery of a far more complete fossil that brings a detailed understanding of its anatomy.

Quetzalcoatlus was named after the feathered serpent of Aztec legend. Its remains were found in locations that were well inland during the Cretaceous, which rules out the piscivorous diet common to many smaller pterosaurs. Instead it is currently perceived as more like a Marabou stork, whose long, pointed beak it shares, pecking at small vertebrates as it hunched around on its four thick and muscular limbs. Those enormous arms created an estimated wingspan of 10–12m (33–39ft). By comparison the biggest flying bird today is the wandering albatross, whose outstretched wings rarely exceed 3.5m (11ft 6in), while the largest ever known is the condor-like giant teratorn, which had wings twice the albatross' length and lived in Argentina 6mya. Whether or not *Quetzalcoatlus* used its gigantic limbs to fly, it was certainly one of the most extraordinary creatures of the Mesozoic era.

Wingspan
10–12m (33–39ft)

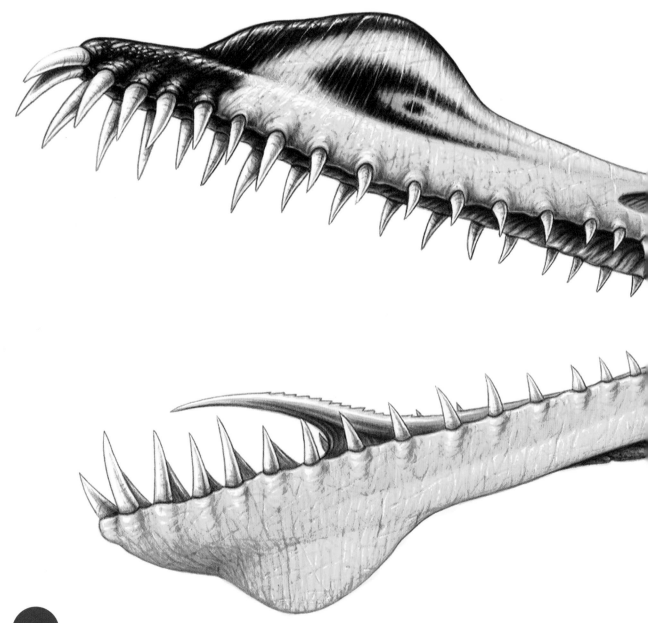

(an-hang-WARE-ah)

ANHANGUERA SANTANAE

Bulging crests above and below the tip of its long beak distinguish *Anhanguera* from most pterosaurs. Its weak legbones suggest it spent most of its life in the air, gliding and flapping its broad leathery wings as it must have been a clumsy walker. As it swooped over the waters of modern South America and Australia, it scooped fish into its powerful bill and impaled them between teeth like curved needles.

CRETACEOUS

MAASTRICHTIAN	
CAMPANIAN	
SANTONIAN	LATE
CONIACIAN	
TURONIAN	
CENOMANIAN	
ALBIAN	
APTIAN	
BARREMIAN	EARLY
HAUTERIVIAN	
VALANGINIAN	
BERRIASIAN	

112–94 mya

PISCIVOROUS

18kg (40lb)

BRAZIL AND AUSTRALIA

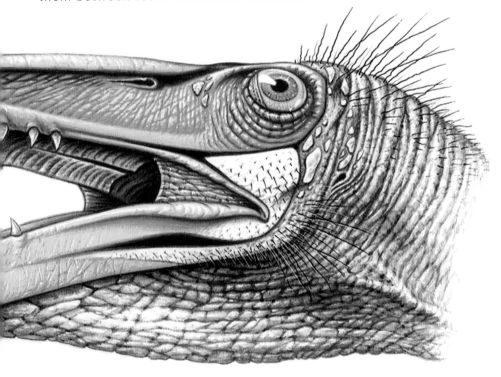

Wingspan 4m (13ft)

MAASTRICHTIAN	
CAMPANIAN	
SANTONIAN	LATE
CONIACIAN	
TURONIAN	
CENOMANIAN	
ALBIAN	
APTIAN	
BARREMIAN	
HAUTERIVIAN	EARLY
VALANGINIAN	
BERRIASIAN	

80mya

C

CARNIVOROUS

2000kg
(1.9 tons)

KANSAS,
USA

(ell-AZ-mo-SORE-uss)

ELASMOSAURUS PLATYURUS

Plesiosaurs were long-necked marine reptiles that thrived through the Triassic to the Cretaceous, and *Elasmosaurus* had the longest neck of any of them, comprising an astonishing 71 vertebrae. By comparison one of the longest-necked dinosaurs, *Mamenchisaurus*, had 19 elongated vertebrae. It was this bizarre beast that so confused Edward Drinker Cope that when publishing its first description in 1869 he mistook the neck for a tail and drew the head at the other end, a significant moment in his feud with Othniel Charles Marsh (see page 89). Cope was working from a fossil found in western Kansas, USA; 80m years earlier these sediments were part of the Western Interior Seaway. *Elasmosaurus* probably pursued schools of fish, tracking their silhouettes by swimming deep beneath them and then ascending with stealth, though there are also indications that elasmosaurids were foragers that reached down to the seafloor to pick up crustaceans. The neck would have been fairly rigid: it couldn't have held its head high above the sea's surface in a swan-like posture, as plesiosaurs – and their modern mythical equivalent, the Loch Ness Monster – are often imagined to do. It gives a name to the Elasmosauridae family, which currently comprises around 20 other plesiosaurs with especially long necks and small heads.

14m
(46ft)

(reh-PEEN-oh-MAYM-us)

REPENOMAMUS GIGANTICUS

1m (3ft)

CRETACEOUS

MAASTRICHTIAN	
CAMPANIAN	
SANTONIAN	LATE
CONIACIAN	
TURONIAN	
CENOMANIAN	
ALBIAN	
APTIAN	
BARREMIAN	
HAUTERIVIAN	EARLY
VALANGINIAN	
BERRIASIAN	

130mya

C

CARNIVOROUS

14kg (30lb)

LIAONING, CHINA

This raccoon-like animal stands out among fellow mammals of the Early Cretaceous because it wasn't a tiny, timorous insectivore – instead, it ate dinosaurs. Two hefty species known to have lived in China – *R. giganticus*, the Cretaceous' largest known mammal, and the smaller *R. robustus* – proved this when one was found fossilised with the bones of a young *Psittacosaurus* in its gut. The bones were barely broken, which suggests *Repenomamus* used its vicious front teeth to gulp down its meals in great chunks.

The remains were discovered in Liaoning province, famed for its feathered dinosaur finds. When these beautifully preserved specimens were described in 2005, they shed as much light on early mammal life as those feathered fossils have on the dinosaur-to-bird transition. At 1m (3ft) long including its tail, with short powerful legs and a 16cm (6in) skull, *R. giganticus* was a squat, muscular predator that could run in short bursts and pounce on its prey. It was part of a family called the triconodonts, which have no direct relatives alive today.

Traditionally, mammals were seen as the poor relations during the Mesozoic era, scurrying about feeding on insects and trying to keep out of dinosaurs' way. *Repenomamus'* discovery shows that they were on the ascent far earlier than previously realised, primed to take centre stage once the dinosaurs had left the scene.

QUIZ ON THE CRETACEOUS

1. What is the meaning of the word 'Cretaceous'?

2. Which two dinosaurs have the shortest names?

3. Which is the biggest feathered dinosaur known, and how long was it?

4. What are the so-called 'raptors' more properly called?

5. Who named *Iguanodon* after the discovery of a few fossilised teeth in a Sussex forest?

6. Which sauropod had a row of tall, forked spines along its back?

7. *Baryonyx*'s discovery in Britain shed light on the true appearance of which huge carnivorous dinosaur?

8. Which sauropod – named by Edward Drinker Cope from a single leg bone – may have been 60m (200ft) long, making it by far the biggest dinosaur ever known?

9. How many melanosomes (pigment cells) could fit across a human hair?

10. Why was *Irritator* given its name?

11. Why was the therizinosaurs' diet unusual among theropods?

12. How are the 'bone-headed' dinosaurs more properly known?

13. Which dinosaur's fossilised skulls may have prompted people to imagine the mythological creature known as the griffin?

14. Which is the only one-fingered dinosaur so far discovered?

15. Which dinosaur is considered to have been the most intelligent?

16. During the Cretaceous an inland sea split North America in two. What is it known as?

17. And what are the western and eastern landmasses either side of it called?

18. Which is the first dinosaur to have gone into space?

19. How fast could *Dromiceiomimus* run?

20. Which dinosaur from Madagascar is the only one definitely known to have been a cannibal?

21. Which sauropod had a neck twice the length of its body?

22. What does palaeontologist Jack Horner think connects *Dracorex*, *Stygimoloch* and *Pachycephalosaurus*?

23. The longest dinosaur name is possessed by one of the smallest-sized genera – which is it?

24. What age could *Tyrannosaurus* live until – and what percentage actually reached this age?

25. The mammal *Repenomamus* is known to have eaten which genus of dinosaur?

For answers see page 340.

DINOSAURS' EXTINCTION

A sky illuminated with fire, an eardrum-bursting roar, then a series of earthquakes that ripped apart the planet's crust and sent tsunamis coursing across the seas: 500m waves raging into the coasts and over the land, destroying all that lay in their way. Volcanoes gushed molten lava down into the valleys, hurricane-force winds spread wildfires around the world... and then there was silence, and years of dust and darkness as a global winter took hold. Any creature that was not drowned or incinerated in the aftermath of the impact would have found itself in a chilling, inhospitable world.

Every 100m years, on average, a huge meteorite collides with the Earth summoning this kind of apocalypse. Around 65.5mya a lump of rock six miles in diameter smashed into what is now Mexico's Yucatan Peninsula, travelling at 20 times the speed of a bullet: one moment it was 12 miles from Earth, a second later it landed with an explosion a billion times greater than the nuclear bomb dropped on Hiroshima. The immediate effect was devastation but its eventual consequence was greater still: to change the course of life on our planet. This meteorite heralded the age of the

mammals and at the same time closed another, that of the non-avian dinosaurs. In 2010 a panel of 41 scientists reviewed 20 years' research and announced a consensus that the meteorite caused what is known as the Cretaceous-Palaeogene extinction event, drawing a line under years of debate. So why did they reach this conclusion? And how do they even know that a meteorite struck the Earth so long ago?

Picture the great craters that we can see on the moon, the result of countless asteroids and comets that battered its arid surface early in its history. Unlike the moon the Earth has an atmosphere, which causes space-rocks to burn as they pass through, often disintegrating before they can land – but still, our planet also bears a multitude of craters. Some are extremely clear, such as the 1.2km-wide (0.7-mile), 200m-deep (650ft) circular hole in the Arizona desert known simply as Meteor Crater, the result of a 50m (165ft) meteorite that landed 50,000 years ago. But others – older, larger impacts – are harder to detect.

In the 1970s geophysicists began searching for oil in the Yucatan Peninsula. In the course of investigations over the years to come, they found three indications that a space-rock of some description had landed here. Drilling a mile below the Earth's crust, engineers found deposits of 'shocked quartz', which is quartz that has had its crystalline structure significantly transformed by exposure to intense pressure. It is only found at the sites of nuclear explosions and meteorite impacts.

On closer observation a 'gravity anomaly' also emerged. Every object exerts a gravitational pull that is proportional to its mass: for instance, a huge object the size of the Earth has a pull strong enough to keep your feet planted on the ground. Smaller things – a pebble, a feather, your body – also exert gravity proportional to their mass, but this is so weak as to be unnoticeable. (You can test this theory by holding a plumb line, a thread with a weight attached to the bottom, next to a massive object and measuring its movement. In most locations the plumb line will descend straight downwards, but if you stand at the foot of a mountain the line will incline slightly towards it.)

This gravitational pull weakens with distance from the object in question and does so at a steady and predictable rate. But in certain rocky places gravity does not act as you would predict. Instead it is stronger or weaker than it should be, chiefly because the rock that is exerting the pull has a greater or lesser density than expected. This is the case at Chicxulub (pronounced *cheek*-shuh-loob), the Mexican town nearest to the site of the meteorite's landing. The impact fractured much of the rock beneath the earth's surface, giving it a lower density and therefore a weaker pull. These variations mean that were you to visit a meteorite impact site, your weight would actually fluctuate a little as you walked around.

The third clue was the presence of tektites in the vicinity. Tektites are small rounded lumps of dark, naturally formed glass created when rocks are subjected to extraordinary degrees of heat and pressure – the kind of furnace-like conditions created by a 6-mile wide burning lump of asteroid crashing into the ground at 43,000mph.

Together, the three factors were proof that something had landed here... but to deduce its size, it was necessary to find the crater. The geophysicists knew that much of it would now be underwater in the Gulf of Mexico but it still seemed strangely difficult to discern any sign of a circular rim.

And then it dawned that perhaps this crater was big. Seriously big. So huge, in fact, that it was impossible to see from the surface of the Earth, and that only advanced surveying techniques would reveal its size. By testing for gravity anomalies and fluctuations in the Earth's magnetic field, they found proof of a submerged crater about 110 miles in diameter, with a suggestion that it lay within an even wider circle of 186 miles across. More evidence emerged at sites all around the world when geologists tested a thin stratum of clay called the Cretaceous–Palaeogene Boundary, which dates from the impact. It turned out to contain 1000 times more iridium than the strata above and below it. Iridium is an element rarely found on Earth – but is very common in meteorites. The impact evidently hurled so much debris into the atmosphere that it came to settle across the planet. If you live in Great Britain, imagine the air clogged with dust from a meteorite landing in Australia.

In fact, it was this huge volume of debris clouding the sky that scientists consider a prime cause of the dinosaurs' extinction. The air hung thick and heavy with dust particles, which blocked much of the sunlight and left the world cold and dark. Plants need sunlight in order to photosynthesise; without it they wilted and died, spelling trouble for herbivorous dinosaurs, which needed to eat huge quantities of vegetation every day to survive. A domino effect through the food-chain imperilled the carnivores that preyed on the herbivores. Dinosaurs failed to adapt to the cold new world with sufficient speed and relatively soon – perhaps within a few thousand years – they became extinct. They were not the only ones to die out at this time: for example, shellfish such as ammonites and belemnites, the pterosaurs and marine plesiosaurs suffered the same fate, among many other species. In all it is thought that 60 per cent of life-forms on Earth disappeared in the Cretaceous–Palaeogene extinction event.

But that means that more than a third of living things somehow endured the apocalypse. Scavengers flourished, living on carrion and detritus such as leaf mulch. Birds were able to fly around the globe in search of food, whereas non-avian dinosaurs could only lumber slowly across the land. Certain reptiles managed to survive: prehistoric crocodiles, for instance, which learned to subsist on occasional meals, a trick retained by their descendants today.

Then there were the mammals. At this time they were mostly very small animals that foraged for food. But in the 20m years (again, a relatively short period in evolutionary history) following the meteorite impact they grew to 1000 times their previous size, a growth spurt propelled by feasting on the vegetation no longer being eaten by dinosaurs. From having in the main been tiny innocuous rodent-like creatures, they became the kings of the new ecosystem. This spectacular development produced some mighty mammals to rival the great beasts of the dinosaur era: for instance *Paraceratherium*, a hornless relative of the modern rhinoceros, which stood 6m (20ft) high – almost twice as tall as an African elephant.

Science does not deal in final verdicts, only in theories drawn from the available evidence. At the time of writing there is broad agreement that the Chicxulub meteorite impact provides the most convincing explanation for the dinosaurs' disappearance. There remains a minority of dissenters, however. Some scientists believe that a period of great volcanic activity in India's Deccan steppes in the Late Cretaceous released enough sulphur into the atmosphere to cause a fatal period of global cooling. Even those experts who

sign up to the massive-impact theory see this volcanic activity as a contributory factor. Others think changes in the Earth's orbit may have caused this cooling effect, still others that a star exploded into a supernova within 10,000 light years of Earth – close enough to expose our planet to lethal amounts of radiation.

Even the date of the dinosaurs' extinction remains uncertain. While the figure of 65.5m years has become familiar to the public, research on a hadrosaur femur in 2011 suggested that the duck-billed dinosaurs may still have been around 64.8mya.

What is certain is the Late Cretaceous marked the relatively abrupt demise of the most terrifying and magnificent creatures ever to roam this planet. That we know about them at all is due to the diligence and dedication of palaeontologists over the last two centuries. It is just as breathtaking that thanks to physicists' and geologists' meticulous research, we now have a good understanding of the catastrophe that consigned them to history.

QUIZ ANSWERS

Chapter One: The Triassic

1. 250mya.
2. There were spines running along it.
3. *Blikanasaurus*.
4. Ornithischians and saurischians, or bird-hipped and lizard-hipped.
5. *Daemonosaurus*.
6. Wales.
7. *Thecodontosaurus*.
8. The phytosaurs.
9. Convergent evolution.
10. *Heleosaurus*.

Chapter Two: The Jurassic

1. Laurasia (the northern continent) and Gondwana.
2. Because their skull bones were so light they rarely fossilised.
3. *Megalosaurus*.
4. *Megalosaurus*, *Iguanodon* and *Hylaeosaurus*.
5. Australia.
6. They had spikes at the tip for self-defence.
7. Othniel Charles Marsh and Edward Drinker Cope.
8. Because its remains were found during the construction of a gasworks.
9. *Deinonychus*.
10. *Giraffatitan*.
11. Twelve.
12. Because it turned out to be the same animal as *Apatosaurus*, which was named two years earlier.
13. Trace fossils.
14. King Edward VII, in 1905.
15. *Pliosaurus funkei*.

Chapter Three: The Cretaceous

1. Chalk period.
2. *Mei* and *Kol*.
3. *Yutyrannus*, 9m (30ft).
4. Dromaeosaurs.
5. Gideon Mantell.
6. *Amargasaurus*.
7. *Spinosaurus*.
8. *Amphicoelias*.
9. A hundred.
10. Because its holotype had been tampered with by poachers who tried to sell it as a pterosaur fossil.
11. They were herbivorous.
12. Pachycephalosaurs.
13. *Protoceratops*.
14. *Linhenykus*.
15. *Troodon*.
16. The Western Interior Seaway.
17. Laramidia and Appalachia.
18. *Maiasaura*, discovered in Montana – an astronaut from that state took a fossil with him on a space shuttle expedition.
19. Up to 50mph.
20. *Majungasaurus*.
21. *Erketu*.
22. He believes they were the same dinosaur but their fossils come from different growth stages.
23. *Micropachycephalosaurus*.
24. 30 years, and 2 per cent.
25. Psittacosaurus.

GLOSSARY

Abelisaurs: a family of large carnivorous bipeds living in Gondwana during the Cretaceous period. They tended to have tiny arms, powerful legs and skulls that were almost as high as they were long.

Alvarezsaurs: small, long-legged fast runners once thought to be the first flightless birds but now considered early maniraptoran dinosaurs. Some are known to have had feathers. They lived from the Late Jurassic to Late Cretaceous in what is now Asia and both American continents.

Ankylosaurs: these squat, armour-plated herbivores lived from the Early Jurassic until the dinosaurs' extinction. Their fossils have been found on every continent but Africa.

Brachiosaurs: huge sauropods characterised by their giraffe-like stance.

Basal: a group, or member of a group, occupying a position at the base of a clade.

Carcharodontosaurs: containing some of the biggest predators ever known, this family lived throughout the Late Jurassic and Cretaceous periods and included *Giganotosaurus* and *Tyrannotitan*.

Clade: from the Ancient Greek for 'branch', as in a branch on the tree of life. A clade is a group of organisms that shares the same, single ancestor.

Coelurosaurs: the clade of often feathered theropods containing the ones that evolved into birds. Most were carnivorous bipeds, and they spanned from *Tyrannosaurus* to the tiny *Microraptor*.

Diplodocids: a kind of sauropod with especially long necks and whip-like tails.

Dromaeosaurs: small to medium-sized feathered carnivores that lived from the Middle Jurassic until the dinosaurs' extinction. The name means 'running lizard'. These bird-like predators include *Velociraptor* and are sometimes informally known as 'raptors'.

Genus: a distinct type of animal, whose varieties are classed as species. The plural is genera.

Gondwana: one of the two supercontinents that formed the dinosaurs' world, the other being Laurasia. Before they separated around 200mya they formed a single supercontinent called Pangaea. Gondwana comprised what is now Africa, Antarctica, Australia, India, South America and the Balkans.

Hadrosaurs: the duck-billed dinosaurs that thrived throughout the Cretaceous period.

Ichthyosaurs: marine reptiles that lived throughout the Mesozoic era.

Laurasia: one of the two supercontinents, formed from what we now know as Asia, Europe (except for the Balkans) and North America; the other is Gondwana.

Maniraptorans: literally 'hand-snatchers', these theropods had long arms, three-fingered hands and, uniquely among dinosaurs, breastbones. Most were feathered – some had soft, downy insulation, but other later species developed flight feathers. Modern birds evolved from these dinosaurs.

Ornithischians: meaning 'bird-hipped', this is one of the two orders of dinosaurs, as classified by 19th-century palaeontologist Harry Seeley. He noted that these dinosaurs had pelvises similar to modern birds, with the pubic bone pointing backwards.

Ornithomimids: the ostrich-like dinosaurs, which included the fastest running varieties known.

Prosauropods: an informal name for the early bipedal sauropodomorphs of the Triassic and Early Jurassic that later evolved into four-legged herbivores such as *Diplodocus*.

Protofeathers: the soft 'dino-fuzz' found on some dinosaurs, particularly coelurosaurs; some genera also developed flight feathers.

Pterosaurs: flying reptiles that lived alongside the dinosaurs throughout the Mesozoic era.

Saurischian: meaning 'lizard-hipped', this is one of the two main orders of dinosaurs. All carnivorous dinosaurs were saurischians. Their pubic bone pointed forward. Modern birds evolved from these lizard-hipped dinosaurs, rather than the bird-hipped.

Sauropodomorphs: the sub-order of long-necked herbivores that spans from the tiny *Saturnalia* of the Late Triassic to the sauropods such as the 30m (90ft) *Argentinosaurus* of the Late Cretaceous.

Early ones were relatively small bipeds, and over time they grew larger and settled on all fours as the biggest animals ever to exist.

Sauropods: with a name meaning 'lizard foot', these are the massive, four-legged, long-necked herbivores such as *Apatosaurus* and *Diplodocus*. The largest species may have weighed almost 90 tonnes (88 tons).

Therizinosaurs: bizarre theropods with huge claws but teeth clearly suited to a herbivorous diet.

Theropods: a sub-order of saurischian dinosaurs whose name means 'beast foot'. All were bipeds, most with serrated teeth and a wishbone. Most were carnivorous but some herbivorous theropods also evolved, in the form of the therizonosaurs.

Titanosaurs: a group of often huge sauropods, such as *Argentinosaurus* and *Paralititan*.

Troodonts: a family of small to medium-sized carnivorous theropods, characterised by their very long legs and a large curved claw on their second toes. They had extremely large brains for dinosaurs, forward-facing eyes and acute hearing.

Tyrannosaurs: the most famous of the carnivores, spanning from *Guanlong* in the Middle Jurassic to *Tyrannosaurus rex* in the Late Cretaceous.

ACKNOWLEDGMENTS

As a small boy in the early 1980s nothing seemed more exciting to me than a visit to the Natural History Museum in London, where I would gaze up at the *Diplodocus* skeleton and later depart clutching some little dinosaur-related memento: a pencil rubber shaped like *Stegosaurus*, a lurid poster of a Jurassic scene, or a book crammed with dino-facts. It would have blown my four-year-old mind to know that one day I'd have the opportunity to write my own book on the subject, and the fact that this has happened owes everything to the following people. I would like to thank:

Rowan, for her love, support and understanding of a husband who sometimes seemed more rooted in the Middle Jurassic than the here-and-now during this book's creation.

Isla, aged four, for the regular interrogations concerning which dinosaur I was writing about and why, and why, and why again, which often helped to focus my mind.

Lottie, aged two, for never failing to make me smile, despite her active nocturnal lifestyle.

Rosa, aged eight months, who arrived in the world as this book's submission deadline loomed but helpfully spent enough of her time asleep to allow me to keep writing.

Mum and Dad, for everything, but in particular running a dino-news cuttings service.

My friend Paul Willetts, my brilliant literary agent Matthew Hamilton and this book's UK publisher Rosemary Davidson at Square Peg and US publisher Matthew Lore at The Experiment, without whom in differing ways I would not have had this opportunity – my sincere gratitude to each of you.

My former colleagues at the *Eastern Daily Press* for granting me a sabbatical in order to begin this book.

This book's appearance owes everything to the following people's hard work: Kristen Harrison and Rowan Powell at The Curved House who oversaw the project, illustrator Fabio Pastori, Simon Rhodes in production at Random House, Jonathan Baker of Seagull Design who typeset the pages, and Sophie Devine who drew the silhouettes.

Stephen Pope and Elizabeth-Anne Wheal, for housing a displaced author in their kitchen for six weeks.

The very helpful Dr Mike Taylor, of Bristol University's Department of Earth Sciences, and Dr Roger Benson, of Cambridge University's Department of Earth Sciences, whose responses to my questions were prompt, kind and generous.

I have striven to produce something accurate from the morass of material extant at any one time. Evidence changes regularly and the same evidence is interpreted differently by different palaeontologists. My thanks go to Dr Darren Naish for ensuring that this book is as accurate as possible at the time of publication. Any errors that remain are mine alone.

FURTHER READING

Dinosaurs: A Field Guide
by Gregory Paul (A&C Black)

Dinosaurs
by Thomas Holtz (Random House)

Dinosaurs
by Steve Brusatte (Quercus Books)

The Dinosauria
Second edition, edited by David Weishampel, Peter Dodson and Halszka Osmólska (University of California Press)

The Great Dinosaur Discoveries by Darren Naish (A&C Black)

The Illustrated Encylopaedia of Dinosaurs
by Dougal Dixon (Lorenz Books)

Planet Dinosaur
by Cavan Scott (BBC Books)

Dinosaurs' footprints are all over the internet... and here are just a few of the many good blogs and websites out there:

Chinleana
Focusing on Triassic palaeontology – http://chinleana.fieldofscience.com

Dave Hone's Archosaur Musings
This British vertebrate palaeontologist's observations on dinosaurs and archosaurs – http://archosaurmusings.wordpress.com

Dinogoss
'Armchair palaeontologist' Matt Martyniuk's views on dinosaur developments – http://dinogoss.blogspot.com

The Dinosaur Society
The homepage of the British organisation for dinosaur enthusiasts – www.dinosaursociety.com

Dinosaur Tracking
Dinosaur news hosted by the *Smithsonian* magazine – http://blogs.smithsonianmag.com/dinosaur

Everything Dinosaur

A child-friendly appraisal of dinosaur developments – http://blog.everythingdinosaur.co.uk

Not Exactly Rocket Science

British science writer Ed Yong's blog often covers news from the world of dinosaur palaeontology – http://phenomena.nationalgeographic.com/blog/not-exactly-rocket-science

The Open Source Paleontologist

Updates on fossil finds from across the animal kingdom – http://openpaleo.blogspot.com

The Paleobiology Database

A comprehensive record of fossil finds maintained by hundreds of palaeontologists – http://fossilworks.org

The Plesiosaur Directory

The latest on marine reptilia of the Mesozoic, from plesiosaurs to pliosaurs – www.plesiosauria.com

Sauropod Vertebrae Picture of the Week

Often shortened to SV-POW!, this is where Mike Taylor, Darren Naish and Matt Wedel share their thoughts on massive Mesozoic herbivores – http://svpow.wordpress.com

The Society of Vertebrate Paleontology

The world's premier palaeontological organisation's website also has much to interest amateur enthusiasts – www.vertpaleo.org

Tetrapod Zoology

British palaeontologist Darren Naish's popular blog discussing news from across the vertebrate world – http://blogs.scientificamerican.com/tetrapod-zoology

All websites are extant and their addresses correct at the time of publication.

MARINE LIFE

Around 90mya in the Late Cretaceous, a dead *Angolatitan* has drifted out to sea and is scavenged by mosasaurs such as the huge *Angolasaurus* and *Tylosaurus*, while a sharp-fanged *Elasmosaurus* plesiosaur also detects a meal. A *Ptychodus* shark and the bony fish *Enchodus* lurk near the seabed as ammonites and *Angolachelys* turtles glide by. Overhead a group of Azhdarchid pterosaurs soar through the sky. All these animals' fossils are known from Turonian stage rock strata.

INDEX